The

Future

of

Prediction

Dr. Don Gregory

ISBN-13: 978-1475217407
ISBN-10: 1475217404

This book is lovingly dedicated to my wife, Lisa, who continues to hang in there through all the chaos, while at the same time working ceaselessly to ensure our comfort and happiness.

Acknowledgements

As of this writing, I have not personally met anyone in the list that follows, but even so, their 'work' has had a profound impact on my learning and thinking. They are all involved with the art of the podcast and my heartfelt appreciation goes to all.

Alan Saunders of "The Philosopher's Zone"
Natasha Mitchell of "All in the Mind"
Ginger Campbell of "The Brain Science Podcast"
Stephen L. Gibson of "Truth Driven Thinking"
... and numerous hosts of "Point of Inquiry", but especially D.J. Grothe, Chris Mooney, Karen Stollznow, and Robert Price

I feel compelled to point out that this book is filled with my ruminations on a number of subjects that I find interesting. You should, however, also be aware that I am not a trained writer (so you shouldn't feel a strong urge to point out the lack of skills thereof), and I am not an academic researcher in neuroscience, nor in the philosophy of mind, so I am sure that some will find plenty of corrections or counter-arguments. I would love to hear from you if you wish to participate in further discussions in a sincere search for better understanding of ourselves and our behaviors.
I can be contacted through my blog at thefutureofprediction.wordpress.com

I would also like to especially thank brother Bill for a stellar job of proofreading and error checking, as well as some insightful comments on the content.

Contents

Introduction
Chapter 1 : The origins of predicting the future … lost in the mists of time
Chapter 2 : Everything we do is about trying to predict the future
Chapter 3 : What does it mean, to predict the future?
Chapter 4 : Why is being able to predict the future important to us, why does it matter?
Chapter 5 : What kind of 'futures' do we want?
Chapter 6 : Why bad news travels fast
    Are there benefits to risk taking?
Chapter 7 : Pseudo-Science / Paranormal
Chapter 8 : Why do we pay so much attention, to 'spiritual' claims?
Chapter 9 : What are the tools & methods for predicting the future?
    Models
        Designed Experiments
        Improving our models
        Levels of theory
        How does the brain develop models?
Chapter 10 : Intelligence
Chapter 11 : Science & Religion
        So, to summarize …
        What is it to "know" or "understand" something?
        Overlapping magesteria
        Intuition
Chapter 12 : Consciousness
    Social Psychology
    Fairness
Chapter 13 : So exactly what *is* the future of prediction?
Notes & References
Appendix 1 : The Statistics of Being a Successful Medium
Appendix 2 : The Justice System and Contra-causal Free Will

Introduction

"I predict" … those are words that capture the attention of anyone who hears them. Those who claim to be able to predict the future have induced awe, respect, faith, hope, distrust, and fear in those around them since the beginnings of civilization. In this book, we will discuss how to predict the future. I predict that if you read through this entire book, you will be fascinated by some of what you read, puzzled perhaps by other parts, and incredulous at still others. I hope that it will be an interesting, hopefully rewarding, and ideally thought-provoking journey for both of us. What could be more fascinating than talking about predicting the future! The discussion has a long and amazing history and, I would predict, a fascinating future as well. We'll start by looking at how prognosticators have plied their trade for as long as people have been around. We will look at some of their tools, methods, practices and results. But I will also look into the implications … what it would mean if we were able to predict the future. I mean, after all, who wouldn't like to be able to do so? Have you really thought about that? What would it mean to you if you could? What would you do with such wonderful capabilities?

We will look into the value to society of soothsayers, oracles, fortune-tellers and other *gifted ones* … but we will go further and will look into the deepest parts of what makes us human – even into looking at how predicting the future is involved in the concept of intelligence ... and even more mystically into the very definitions of consciousness that seems to mark humans as being

1

different from every other creature on Earth. I will even suggest that predicting the future predates human beings ... what an interesting concept ... if not humans, then *who* was there to be making predictions? We will explore this idea if you remain interested.

There are many associated topics to predicting the future, and still many mysteries out there ... perhaps none more so than science's marvel at how the mind can affect our health; some call it the placebo effect; others in history might very well have called it 'a miracle'. There's the practical question of how we decide who to believe, who to depend on for their understandings and abilities ... who will help us predict what will happen to us. Who and what do *you* believe, and most importantly ... why? We will go over these ideas and I will make a series of predictions myself. I will predict that of those who read through this work ... many will disagree with some (or most!) of what they read. To those, I welcome your own thoughts; but I also challenge you to offer your own ideas and your own evidence. I also predict that many people will know a great deal more about these subjects than I do. To those, I welcome your knowledge, your suggestions and criticisms. I look forward, in the future, to learning a great deal more than I know now, about many of these subjects through your feedback and ideas; I hope it will be as interesting for you as it has been so far for me. And I predict, it will continue to be an amazing journey into future.

Chapter 1 : The origins of predicting the future ... lost in the mists of time

So let's talk about predicting the future. How long has it been going on? It's hard to say when predicting began. But I suppose we can easily imagine early humans staring into the sky and wondering if the patterns they seemed to see in the stars meant anything, and whether or not there were unseen forces or beings that, if only they'd look down upon the earth favorably, would ease all the hardships. It's easy to imagine a situation where someone remembered getting angry just before a storm arose, and wondered if there was a connection. Today we recognize that we humans almost obsess on making stories out of the world around us; 'building a narrative' as social scientists call it. It's a way of trying to figure out the *cause & effect* of things, so we can better remember important events and use them to expect what's likely to happen next.

There are many books that talk about the history of prognostication, mentioning the Oracle at Delphi ... others ... so we won't go into them here.[1.1] But have you ever given thought to whether or not prediction even preceded humans? Did (and *do*?) other animals try to form links between their actions, and whether or not things would work out better? Surely they do today; one of my favorite examples is an experiment performed on birds[1.2]. Each morning, on a regular schedule (note, meaning 'predictably'), scrub jays were placed in one of two separate areas. One area had food, the other didn't. Each evening, the birds had access to plenty of food and both of the 'breakfast' rooms. What the researchers found was the jays would hide food in the breakfast room that would not have any the next morning, far more often than in the room where they had learned they could expect food the following day. The scientists suggested this indicates an awareness of future events and planning in the jays; i.e. predicting and acting so as to improve their future circumstances. So predicting may be a common feature in nature, but surely humans have truly made it an art form.

In fact, humans' ability to predict the future better than the rest of the animal kingdom might very well be what differentiates us from the rest. After all, many tests of intelligence are really about being able to accumulate, retain and use information in novel ways. This is exactly what the brain is doing all the time, and I suggest the whole importance of that is to improve our abilities to predict our future. One common intelligence test for animals involves delayed-gratification, where an animal is placed behind a clear plastic screen where it can see food on the other side. The wall is built so that the animal has to backtrack around the wall, moving away from the food, in order to get around and have its treat. How far the animal is willing (or capable) of moving away from the food is seen as an indication of its ability to keep the plan in mind, and continue to work longer and longer distances and times, to have its reward. This is "one" measure of intelligence, and is intimately connected with the animal's ability to remember, plan and predict.

Scientists have also looked at the genomes of humans and many closely related species such as monkeys and apes. Everyone

4

knows that the genetic makeup of chimpanzees is very close to that of Homo sapiens. But everyone equally knows that there is a vast difference in the mental abilities between 'us' and 'them' (and I will suggest primarily in the ability to predict). Can that obvious difference be backed up by science? Some scientists have suggested that the difference has to do with the very parts of the brain that map to the planning and imagination capabilities which humans possess. Scientists have found a relatively small number of genes that appear to be unique to humans that are involved in neural development. They refer to these genes as "Human Accelerated Regions" (of the genome) or HARs.[13] Because many of these particular genes are known to play a role in neurodevelopment, it is suggested that they may affect the areas of the brain that are responsible for language and complex thought which separate humans from other species.

So, maybe there **is** something about brain function and capacity that makes us different from other animals. I would suggest (we will talk more about his later) that this capability is fundamental to better predicting the future; humans excel in this, and it has benefited us all greatly. But is it all about the brain? How does the brain *predict the future*? Do some brains 'commune with the universe' in ways better than the rest of us? How would that work? Can any of us learn to predict the future better?

Historians acknowledge that prophesy, prognostication, divination, all kinds and manners of predicting the future, have been going on since early mentions in recorded history. And it is still important

5

to some even today (yes, there are people who **really do** refer to their horoscopes before making important decisions.) It's clear that predicting the future is important to many people. But is there really anything to it? If it was a complete and total sham, would these practices have been able to survive for thousands of years? Scientists say that if we were to keep track of predictions, both when they seem to 'hit the mark' and when they don't, the analysis would show that there is really nothing there for us to depend upon; but still many people do. Can it not be real? The important question I would have for those scientists is "If this isn't for real, why has it been around for so long, why do so many people take it so seriously?" Some scientists have taken that challenge and have talked about the brain's need to find patterns to enhance our procreation and survival. They also talk about the need to create cause & effect relationships so that if similar events happen again, it may be possible to anticipate what will happen next. All of this is for the purpose of surviving (or surviving **better**) and procreating. Surely if someone were to have developed some ability to know what was going to happen next, wouldn't that be a tremendous adaptive advantage? A Dartmouth paleobiologist recently suggested that one of the most amazing phenomena in evolutionary history, the Cambrian explosion, had to do with species needing to rapidly change (in geological time) because for the first time animals began pursuing and eating each other.[1.4] I would also suggest that the larger neural capabilities that humans acquired became more and more important when humans began trying to predict each other's actions. (I think I know what you think I'm thinking.)

So, perhaps we can agree that trying to predict the future really is important. Even the most logical, rational, methodical scientists among us, were they able to have a high confidence about what was going to happen tomorrow (lottery, horse-race, world-series victor etc.) would definitely take advantage of the information. But, let's not get going too fast too quickly. We haven't really talked about what 'predicting the future' really means. Let's look at some details, just so we're all on the same page here, before we go forward.

Chapter 2 : Everything we do is about trying to predict the future

Let's start by making sure we are all talking about the same thing. Predicting the future, I'm sure you'll agree is something we all do already. Each time we put our foot down our brains are predicting where it will land (for those who didn't see that step-down at the curb and got a **jolt** when you dropped that perilous 4 inches, you'll know what I mean.) Our brains and bodies, acting in unison, are constantly taking care of the next few seconds for us; reaching for the coffee cup as we're reading a good book; reclining in a chair expecting to hit the back of it (and sometimes falling unceremoniously when we're mistaken); and running after a football a friend has passed to us. But it's true in the longer sense too. Why do we go to work every day? To earn a living. Why? So we can use our pay to improve our future and the futures of those we care about.

Why do we go to school, read the news, talk to friends … why do we do everything we do (everything!)? It is about learning how the world works, learning how other people behave and expect us to behave, and learning how things 'go together', both in space and in time. Where is my next meal coming from? How can I ensure the safety of my family? Should I use some money to buy a new car, or invest it instead for a more secure future? What will happen if I hold onto this firecracker and light it?

All of this learning (much more on this later) is for collecting information so that in the future, we might retrieve it when it could be handy. We are also constantly replaying scenarios of events in our brains, comparing memories with currently unfolding surroundings to try to anticipate what will happen next. We do this not only for the next few seconds (step down off the curb), but also in an attempt to make plans for the coming days, weeks, months, and years.

Everything we do and think about is for the purpose of figuring out how we can increase the likelihood that we and our friends will benefit (more on this later in a section talking about something called YESIR!). One of the hallmarks of human relationships is trust and dependability, which reflects our ability to predict what others will do. The underlying idea is that better knowing this will benefit us individually and/or benefit our family and friends.

So, prediction is second nature to us all. Some have paid particular attention to their abilities, honed them, practiced them, and offer them for service (and a fee of course). Some say they use nature itself (mediums), and others use even more mysterious recipes that seem anything but natural (we call them statisticians and insurance under-writers). Many people are willing to pay handsomely to listen to these prognosticators, hoping that what they see in their tea-leaves, or printouts, can be used to make them rich and famous. Before we look at each of these techniques, let's talk about what it actually means to predict the future.

Chapter 3 : What does it mean, to predict the future?

It seems obvious what we mean by predicting the future, doesn't it? But let's make sure we're really focusing on the *kind* of 'predicting the future' that we're talking about. Some kinds of predictions are easy. For example, in a kind of ridiculous way there are situations where I suggest I can predict the future with 100% accuracy. "On the next trading day, the stock market will either go up, go down, or stay the same." One of those scenarios is bound to be correct. Ok ... I told you it was ridiculous, but this simple example can also appear at times when it's not so obvious. "The missing child will be found near a body of water." Do you know how much water there is out there, and how easy it is to be *near* some water somewhere? It isn't guaranteed that a prediction like this will be correct, but the chances are pretty likely. There are plenty of other examples of predicting that are similarly simple, but can also pop up at times in ways that aren't so obvious. Getting through these obvious and simple examples of predicting to get to what we're really after is perhaps a good thing to do. Let's look at some other examples.

In that first case I kind of cheated by making a prediction that couldn't possibly turn out false. How about this one? Suppose I make a prediction that is very very likely to happen, but not **absolutely** certain. Would that be interesting? How about this one? "Tomorrow the sun will rise." Ah, I see, your response is, "yawn." So, you're saying it's not enough **just** to say a prediction will be correct, because it may either be like the first example, or something so obvious or agreed upon that we don't care if someone is predicting the sun will rise. Dare I say it's not **useful** to us, as we bumble our way through our busy lives, to ponder whether or not the sun is likely to rise tomorrow. But, suppose you were living in an area that is under a severe drought and someone comes along who tells you (for a price of course) that he can "definitely, most assuredly" predict whether or not it will rain tomorrow ... and even better yet, he tells you that he has some **control** over whether or not that prediction will come true. How much would you be willing to pay to have rain? So you pay up. (By the way, if this sounds too contrived ... try substituting 'cure for cancer', and assume you or a loved-one is currently stricken ... that maybe someone is offering something to do with the newly discovered, pharmaceutically-suppressed Amazonian super-duper-plant root ... how much would you pay?) So our rain-diviner takes your money, and tells you that there will definitely be rain tomorrow. What would your emotional response be (or think about the cancer cure). Relief? Gratitude?

Then of course, the next day comes and goes without a cloud in the sky. **Now** what would your emotions be? (Are you catching on to the fact that emotions come into play here a lot?) In fact, let's suppose you do find the shyster again and you, in all your emotional fervor, point out that you paid him well for his prediction. And indeed, he shows you a weather web-site map, dated yesterday, but picturing lower Slobovia where it did, in fact, rain. He was right. He predicted rain, and it rained. You've been scammed (and that's why you'd probably not be able to catch up with him again once he takes your money ... happens every

11

day.) Again, this is another simple example ... but surely you know that this scenario plays out many many times every day in more complex ways. And, it doesn't even have to be a sham ... very sincere, very well-meaning efforts sometimes make predictions that turn out not to be true. We've all heard of some kind of claim of some kind of danger that turns out to not be as clear as originally thought. Some examples were studies in the 70s that appeared to find a link between electrical power-lines and leukemia ... and disease-causing silicone breast implants ... and I won't even get into the still-ongoing hot debate about cell-phones causing brain cancers. (And, please, I'm not really advocating one side or the other ... I am just saying that these are cases where the original alarm has had counter arguments raised.) The point being ... how is a mere mortal to know? (We'll get back to this question later on.)

It seems that predicting something obvious, or something very expected, isn't really what we're interested in. Is what is important the ability to predict things that are unexpected? Would you be impressed if, on the day that a famous celebrity figure is caught in some scandalous situation, I showed you a web-site from yesterday, where I predicted just that occurance? (oh wait, maybe that's not all that unexpected ...)

Of course, perhaps I had some inside knowledge about that event; but those who know me also know I'm not in the inner circle of any celebrities or any high-society gossip columnist. Would you be impressed at my, apparent, successful prediction? Probably not once I showed you the several hundred other previous postings, each day for the past year or two, predicting the same thing about the same celebrity. Playing the odds, perhaps I felt that sooner or later it was bound to come true. And when it did, I hailed to the world my amazing abilities! The point is, if you keep predicting the same, unusual, unexpected event, perhaps at some point it really will happen, and if you didn't know about my thousands of 'misses', the eventual 'hit' would be impressive. Again, I won't mention anything about any current *mediums* in the public eye, or, heaven-forbid, any doomsday predictions. Speaking of some (in)famous people making predictions in the popular media ... what if someone is good at predicting things, but in an area far removed from your interests. Would you trust them to predict **your** future; would you hand over your paycheck? Here's an example. Once I was asked what would happen if a proton were brought near a phenylalanine ring; where would it bind? (I predict this is a subject that is pretty far removed from your interests.) Lucky for me, I had just studied an example of this not too long before, and the University President who posed the question to me was amazed I knew the right answer. So, I had developed some skills in predicting molecular interactions. Who cares? Knowing that, in some (relatively) obscure area of knowledge I have made some successful predictions in the past, would you then pay me to *see* your illness and predict the steps you need to take for a cure? Know of anyone like that? Do I need to mention celebrities and their medical advice? hmmmm

So to summarize so far, we're not interested in predicting the obvious, and we have to look very carefully when someone offers a prediction. What I suggest we are **really** interested in is what makes a prediction **useful** to us. Ultimately, it comes down to the **likelihood** of the prediction; the less likely a prediction coming true, and then actually coming true, the more useful or valuable that prediction becomes. So how do we understand more about what things are likely to happen?

The study of probability helps answer the 'misbehaving celebrity' example given above. A rare probability repeated over and over again adds up the individual probabilities. This means that, given enough repetition, a rare event becomes more and more probable, or *less unexpected*. Another view on this is if I ask, "What is the probability (each day) when I predict a particular celebrity will get into trouble on that day?" For each separate day it's a rare expectation. But if today I ask, "what is the likelihood that any specific celebrity will get into trouble sometime in the next 10 years?" then the likelihood goes up a lot (my apologies to all you innocent celebrities out there of whom this is not true). This is a subtle situation that can creep into thoughts about prediction. Suppose I were to predict that a green pickup-truck, with a license place that has either a 'B' or a 'P' in the third position, and that the last digit of the license place is a 9, would appear at a specific intersection in a particular city on the other side of the country, driven by a red-headed woman, with a small white dog on her lap and with a woman in a orange blouse looking on from the curb? What are the odds of that happening? That would be a very, very rare occurrence. The very next day, I produce a photograph of

14

just that situation. You've already seen through the scam here haven't you? Of course what **seems** to be extremely rare, with odds of it happening being 1 in a kazillion (or more) ... is ridiculous when you know that this "unimaginably small likelihood" had already happened, and I just described what had already taken place. Does **this** situation ever happen? I would just mention a somewhat mysterious phrase here .... cosmological constants.

The importance of a prediction is also about how we, individually, feel a prediction will impact us (i.e. how useful is it to *us*?). What if I predict that every time an ant walks up to a lump of sugar, it will circle it in a clockwise direction (hopefully it's clear I just made that up, I have no idea ...). But, let's suppose I'm right. Who cares? You probably wouldn't, so while I might be correct, and while it might be useful to someone writing their PhD thesis on 'ant decision making', for most of us, it wouldn't be very useful "to us". So the concept of "useful", and more importantly, "useful-to-us" ... is a large part of what makes a prediction valuable.

So, exactly what is "useful to us"? Well, of course, that is different for each of us, and it is one of those things that is part of the accumulation of our life's experiences, what we wish our future to be, and how things are going for us in our current world circumstances. We'll discuss some important social aspects of prediction later on. But let's briefly consider the social impact of soothsayers, seers etc. First, dispense with the obvious ... *if* someone could truly predict the future, of course we'd all know about it. They'd be famous and most likely rich & powerful. There have been several people in history who have pronounced their abilities to predict the future. Some of those have become rich and famous. They were real, right? But even if it weren't true (that they really had *the gift*) can there possibly be a social value to the village fortune-teller regardless of whether or not they are right or wrong? For all of you rational, scientific, objective, logical readers, ask yourself why these kinds of people have been around throughout history, in all societies. No doubt if they were a complete sham, they would be chased out of the village (where everyone knows everyone else). It's true there is always the explanation of how humans want to establish a cause & effect narrative, to recognize patterns and to "make sense" of a chaotic world. But I would suggest there is a deeper value that has been overlooked. Obviously if someone set up shop and started charging people, telling them they were going to win the lottery the next day his career would be short lived. But we all know in reality it is not quite like that. Such predictions are more subtle ... "danger in your future" ... "opportunity if you are only bold enough to embrace it". Even the most skeptical admit that those who practice "the art" today (and no doubt in the past) were shrewd observers of both people and the general social environment. Imagine if you will, someone who made the effort to keep track of the current gossip, who possessed excellent social skills to draw people out in casual conversation, who had a knack of being able to piece together bits and pieces of information that he picked up, in order to draw some useful conclusions about what

16

is going on around him. (Sounds a bit like a modern-day market analyst doesn't it?) The best of these people could either be trusted advisors to people in power, but for whatever reason, they might also "set up shop" and tell people things that were likely to be true, based on their information gathering skills. Imagine the following interaction. Blacksmith Ben is worried. He thinks his daughter is spending too much time with that horse-stable kid and he wants to know if anything bad will come of it. Sally Soothsayer, of course, has heard the local gossip, and has observed that actually, the daughter isn't interested in the stable-hand, but that she and the wealthy son of the horse ranch owner seem to give those secret glances at each other whenever they pass on the street. So, she sets Ben up by saying she sees possible danger for his daughter, but that through her contact with the masters of the universe, she can change things so that his daughter will have a change of heart and perhaps will marry into a better social position after all. But, for this to happen, of course, Ben will need to pay Sally for the emotional and physical drain of the séance, asking the masters for their help; the more Ben is willing to pay, the more likely the masters will see the importance to him. So the next day Ben returns with a couple of wheelbarrows of precious stones he has collected from his blacksmithing over the last couple of decades. In the meantime, Sally has sought out the wealthy son, and casually mentions that she has heard that the blacksmith's daughter is thinking of going off to business school on the other side of the country, but … with a wink … she also tells him that she overheard the daughter mention to her friends that if she were only to have a stable home with a wonderful husband, she could happily spend the rest of her life in the sleepy village. When Sally collects Ben's retirement savings (sounds ludicrous doesn't it, could never happen in *real* life.) Sally beams and tells him the masters have decided to help, and they have suggested that it's very likely (leaving herself the all important 'out') that the daughter will, in fact, marry the rich son.

There are many silly but subtly noxious aspects to this story. The main point is the social value that Sally brought to the table. She of course benefited from her toils (precious stones are always welcome), but when the daughter married the rich ranch son, they and her father were thrilled. You can see how Sally's reputation would grow and others would come to her with their wheelbarrows. Similar stories could be told where, if only the town mayor would approve that bridge to be built, he would avoid that near-death sickness that Sally sees in his 'aura'. When the bridge is actually built, something that Sally *saw* would be welcomed by local business owners, and the town prospers, **and** the mayor indeed avoids the plague ... Sally could continue her social engineering by reminding the mayor of her role in the good that befell the town, and the bad that she helped him personally avoid. Think these are far-fetched little fairy tales? Think again. In fact, there is an interesting analysis exercise we will go through later on, showing how likely it is that someone like Sally can prosper in making wise, subtle, predictions. Sally also did a couple of other critical things. Although these seemed to be second nature to her, others (we call them academicians) have made a

18

study out of how these things work. They have found that if you collect important, relevant information and continue to "refine your model" (In this case Sally's keen sense of how people behave and what's important to them) and seek those things that can be modified to affect the outcome (who is influenced by whom etc.) then you not only have a good chance of being able to predict likely outcomes, but perhaps even more important, **change** those outcomes to suit your preferences. Lots more on this later.

But for now, at least let's agree that when someone makes a prediction it can sometimes be a bit tricky to know what to make of it. Is the prediction *unexpected* (especially if it involves something that is outside of our regular experiences)? And who is making the prediction? Would the prognosticator somehow benefit (financially or otherwise)? Even if not, do we know anything about their record of hits and misses? Could it be a scam even though it seems legitimate? Surely this is one of those areas where things are not as simple, and as easy, as it might first appear. Perhaps a wise person once said it best ... "Predicting the future is easy, but getting it right is almost impossible!"[3.1]

Chapter 4 : Why is being able to predict the future important to us, why does it matter?

It's one thing to know how the future might turn out (or at least have a good guess). But, as we mentioned previously, it's an entirely different thing in being able to **change** the future. There is, however, another more subtle value to being able to make accurate predictions. It has to do with some of our most basic social needs. Key to this is the fact that we each want to be the *special* person in our group, someone who is admired and respected. This is true whether it's because we are admired if we can plan and execute the best dinner party, or whether we have the best piercings and tattoos amongst our gang. Teen-agers (and we adults too) spend great effort to "fit in", even if it's to a fringe group that is far in the minority (in fact, often it's even more attractive to be *special* in a small rebel group). So it's understandable that some people who, for whatever reason, don't seem to bubble up to the top of the "popular group", drift to more unique social niches It's interesting to note the relationship between 'seers' and their being social outcasts, either personally, or as part of a group, even as a whole culture. We can easily see how being perceived as having special knowledge or access to the masters, could bring that attention and respect from others. If someone has *the gift* about what the future holds, more so than other people, she can provide that useful knowledge for a fee, and in return, she would be more likely to be seen as special.

You have no doubt heard the saying, "knowledge is power." But power to do what? Scientists gain great respect by discovering (or creating) a new understanding of nature. What this really means is that they describe a new way to predict what will happen in nature if, when they do 'A' then 'B' will happen. This can be very powerful for society, but also for the individual. Power more often means the ability to influence and/or command other people. One interesting aspect to this is that **if** someone were able to predict the future, it would be to his benefit if not everyone were able to do so as well. If that were the case, he would no longer be 'special' and so could not trade his gift in exchange for benefits from others.

Throughout history, those that purported to be able to predict the future have been given great prestige and power by those in charge, in return for predicting the future, but just for them only (those in charge). Just as significantly those leaders/rulers-in-charge have made an effort to outlaw, or at least restrict, predictions being made by those not in their control. They want that power for themselves. Kings used to put to death any but their own soothsayers. And it's not just in the dim past that fortune-telling and witchcraft (which can be seen as attempts to affect the future) had fatal consequences. It happens even today, and not just in remote villages. It happened not too long ago in London.[4.1] It's only been very recently[4.2] that the State of Vermont removed criminal punishment of convicted soothsayers. In 1956 a Helen Duncan was arrested in the UK for "pretending to communicate with spirits". In 1929 New York State amended its witchcraft laws to "exempt ministers and mediums of Spiritualist associations acting in good faith without personal fees.[4.3]"

We value, and fear, those who claim they can confidently predict the future. We all would like to be able to do so, but since we usually feel relatively helpless as the world spins around us, we *fear* the future because we feel we have little control over it. If we were only able to understand how things work, how to see danger better, how to take advantage of opportunities better, we would be less fearful. We have evolved to look for patterns, for cause & effect, and to seek narrative-forming information (even inventing new narratives if necessary). We have the ability to do so better than other animals out there in nature. However, when we don't see those patterns clearly, we become fearful since it seems more likely that nature will overwhelm us instead.

We have mentioned how our interactions with others, and our ability to predict their future actions, has perhaps been the most important step in human evolution. The ability to predict is even more important than that. As we shall see later, it goes to the very aspect of our having consciousness. So, is it important to us? It is *the* most important thing you do. Could you do it better? Actually, whether you think about it often or not, you are working on that every day, all the time. So, let's look more closely about what it means, to be able to predict the future better.

Chapter 5 : What kind of 'futures' do we want?

It's important to keep in mind that it is not just about "knowing" the future, we actually want to **control** the future. It is an easy choice. If it were between simply knowing that something bad would happen, and being able to do something to avoid it, clearly we would take the option of being able to change a bad outcome. So again, let's dispense with the obvious. If someone truly had *access to the masters*, or to the *spirits beyond* and would be willing to offer their help, that would be great. Unfortunately, throughout history, there have been far more frauds that claimed this ability, than people who sincerely believed they did. So what other options are there? Can we all predict the future? Can we improve how well we can do so?

We discussed some obvious, trivial, kinds of future predictions earlier (the mind anticipating stepping off the curb for example). There is powerful subtlety in thinking about how this works, understanding it, and trying to improve our abilities to be able to predict the future better. If we're not lucky enough to have a direct line for cosmic revelation, then we should look at the techniques, tools and methods that we **do** have access to. We briefly mentioned these ideas before; to better predict the future we need a good model/theory/idea of how things work, and enough information to plug into that idea to see how well we can do in making predictions. Of course scientists, mathematicians, and statisticians have been working in this area for many decades.

Models and errors

How do we improve our ability to predict the future? Perhaps the best way is to check how well things actually work out, and see if we can find the reasons we may make errors. Everyone makes mistakes, so the goal would be not only to understand how mistakes arise, but also to reduce them if possible. Once the source of the mistakes are better understood, perhaps something can be done to decrease them; maybe this is improving our idea/theory, perhaps it's improving the quality of the information we're using. Mathematicians use the terms "model" for the idea or theory, and "data" for the information, but it's still the same concept. Every day, we go through life with our internal **model** of how the world works. From early in development, our brains have some pre-built models that are hard-wired (newborns for example, will instinctively recoil from images of snakes, and will become terrified if they're placed high up in a room, even if they're perfectly safe). Even in the womb, babies begin to accumulate experiences and record them in their brains.

One of the most important points to be made in this book is about how the brain accomplishes this astonishing feat. How we use acquired knowledge to improve our own models, to better predict the future, perhaps to be able to *change* (!) the future, to improve the lives of our loved ones and ourselves. However, we first need to discuss some other topics about prediction, so that we can better understand how our brains use the information we've accumulated throughout our lives.

24

Chapter 6 : Why bad news travels fast

We have mentioned the desire to be able to change the future. Assume we're currently in a safe situation; any news that predicts how we could improve our situation, is well received. But **everything** has a risk, and we intuitively know that. Even acting on 'good news' takes energy, or time, or other resources, so has a risk of wasting those resources if the good news turns out to be wrong.

So "a bird in the hand, is worth two in the bush," and scientists who have tried to study this[6.1] (showing the wisdom in old-sayings) note that people value something more highly when it is something that can be lost, as opposed to when it is potential gain. Generally, the preference to keep what you have rather than risk it for a potential greater-good, is a factor of 2 to 2.5. The status quo is given a lot of psychological and emotional weight even if there is a potential for possible improvement.

In the work these scientists did, there are descriptions of two experiments. In one, half of a group of subjects were each given a coffee mug. Then, those who got a mug were asked the price at which they were willing to sell it, and those who didn't get a mug were asked what price they were willing to offer for one. Historically, economists and social scientists predicted that both prices will be about the same if everyone were thinking and acting rationally, since the coffee mug should have the same value for everyone. But in fact, the desired selling price was typically over twice what the other group was willing to pay. The subjects valued something they had 'in

25

hand' (literally) twice as much as those who didn't have it.

In another experiment, subjects were each given either a pen or a mug with a college logo, and were told that both were of roughly equal value. Then the subjects were offered the opportunity to exchange the item that they had received for the other. If the subjects' preferences had nothing to do with the item they received, the fraction of subjects keeping a mug should equal the fraction of subjects exchanging a pen for a mug, and the fraction of subjects keeping a pen should equal the fraction of subjects exchanging a mug for a pen. In fact, most people kept the item they received, regardless if it was the pen or the mug; only 22% of subjects traded. Most people will reject an even-chance gamble (50% of winning, and 50% of losing) unless the possible win is at least twice the value of the possible loss.

This is all about the study of risk, and there are many people who make a living assessing risk to benefit values (i.e. they're *predicting* the future value of things and actions). There is an excellent discussion in a book by Ropeik and Gray[62] that describes some of the psychological aspects of how people view risks in loss & gain.

Familiarity breeds acceptance
Most people are more afraid of risks that are new than those they've lived with for a while. Fear of the West Nile virus declined between 1999 and 2001, even though the number of cases was still significant.

Nature is good, mankind is evil
People have been fearful of radiation risks from nuclear energy for decades, and recently even from cell phones, but are far less nervous about the most dangerous source of radiation known, one that causes

way too many deaths each year, the sun.[6.3]

I'm in control
Similarly, someone who smokes may not buy a house with asbestos tiles (and perhaps they shouldn't, but the point is they are already doing damage). But, smoking is also known to reduce stress … so …

Risk/benefit analysis
People will also choose to live in places with known dangers such as earthquakes or tornados. A huge percentage of the earth's population lives near a coast, on the assumption that hurricanes and tsunamis will happen elsewhere.

The ick factor
Fear is greater, if the injury is more **awful**. People fear lightening strikes, and shark attacks far more than heart-disease (again, perhaps some sense of control comes into play here.)

Reputation comes into play
"Trust" (and we'll have more to say on this later); suppose you were given two glasses of a liquid to drink. It "looks" like water. One is offered by your best friend, and one is from that junk yard down the street. Which would **you** be more likely to drink?

Risks can be seen as somehow 'relative'
According to the authors, fear of terrorism was very high late in 2001 (seems reasonable), but fears of street crime and other more common risks were down even though the risks were not.

The less we're certain, the more we perceive risk
And perhaps this is correct, uncertainly maybe *should* be seen as more risky; however, uncertainly simply means that there could be big benefits, as well as big losses. But since we don't value those equally, we tend to be more averse to those things we're less sure of, such as many new technologies.

Even genetics comes into play
Parents are way less comfortable with their children doing exactly the same things they did as kids. But even aside from the obvious, parents are also more concerned about the dangers to their children, even if they face those same dangers; the authors described differences of concern between asbestos in a school, *vs.* in the parent's workplace.

It can get personal
People will give greater attention to risks when they look like they will be directly affected. This also makes sense in the context of risk/benefit. Taking action on a risk that affects others takes

resources. But if you stand to lose something important, then the resources will be more warranted. There had been wars raging around the globe, and terrorist attacks before 9/11 … but afterwards, things in America changed dramatically.

Are there benefits to risk taking?

Note however, risk-taking can be advantageous to others in that risk-takers, while often failing, will sometimes discover a benefit. If they share the good news, or the windfall, with their 'friends & family' then everyone benefits. For others who allow risk-takers to take risks, if the end result is failure, then (generally) only the risk-taker suffers. So, risk takers (i.e. adventurers, heroes) are given great status and admiration even by those who would never be willing to take those risks themselves. Many are willing to support, both monetarily and socially, risk-takers if it means it is more likely that those who succeed will share their discoveries. Some have suggested this is the evolutionary advantage to risk-taking, and might even partially explain hyper-active kids; they're more ready to act before thinking.

However, bad outcomes can be a significant negative factor against a comfortable *status quo*. So while one might take action on good news, bad news almost always deserves high attention and generally a commitment of resources. We are more alert, and more sensitive, to bad news. We all often lament the constant barrage of negative news and the 'car crash' mentality of today's newscasts – but note our lamentations are most often expressed while we're watching the same. We place special emphasis and attention when something bad happens. It is an interesting

28

observation[6.4] that people will stop to wonder why something bad happens, but usually not when something good happens.

We don't often ask *why* a man threw himself on top of a stranger who fell into the path of an oncoming subway. We don't ask *why* a hurricane season wasn't nearly as bad as predicted. We tend to focus upon, and retain recollections of, negative experiences. This helps explain the phenomenon of being fairly certain that the other checkout lane, or the other traffic lane, almost always seems to be moving more quickly than our own. When ours is the faster moving, we don't focus upon that.

So, if we see a positive future, we probably won't want to change things very much. But the opposite is very true. If we see a negative future, we definitely would want to change it. It's a matter of control. Even if there are tough times ahead, we're much more comfortable if we feel that we have some control over unfolding event. A sense of control (or lack of it) is why so many people are more afraid of flying than driving, even when it is clear that safety records warrant just the opposite.

There's also the need by those who provide information (gossiping neighbors, the media etc) to find ways to make their information appear more important, so that they, themselves, increase in perceived value. Even respected scientific literature, with the best of intentions in mind, often helps muddy the waters. One example is a short paragraph in the issue of Science News, 1/14/2012, p. 9 describing differences in parasitic infections in two groups of rural school-aged children. The section is titled "School doesn't bug some kids", and the text of the article focuses on the whether the difference in rates of infections between two groups of children are from whether they are home-schooled or attend a public school. Less emphasized however, is a brief note in the text that the most probable source of the observed differences in infection is the time that the home-schooled children actually spend working in their farms' irrigation canals. The canals are where the infectious parasitic worms live. The researchers appear to have been collecting information with a focus on the children's schooling methods. But by emphasizing this in the title and in the article text, the authors (inadvertently?) allow the possibility that readers will make an erroneous connection between the type of schooling and the rates of infection, when in fact it is more likely that the time spent in the canals is the more important factor. **If** the researchers were to have compared the time spent in the canals and the rates of infection, I predict that they would have found that there is no additional significant impact to be had by including where the children had been schooled.

In all of this discussion about how to assess risks and benefits, we haven't yet talked about how people have tried to deal with determining what's about to happen to them. We've mentioned access to the spirit world, and we've mentioned academic mathematical gymnastics such as statistics. But let's look into these a bit more deeply, to understand not only what they're about, but why some people value some of these methods more highly than others. We've talked about how risks and benefits aren't considered equally. So it would make sense that, if we weren't very confident in our own abilities to see bad news coming, we might turn to others who claim that they can. Let's talk about those claims.

Chapter 7 : Pseudo-Science / Paranormal

There is not a whole lot one can say about *claims of the paranormal*. Either it's real or it isn't. But what we're talking about here is how we can better predict the future. In one sense, it doesn't matter if spiritualists are for real or not. What really matters to us is, can we **depend** on their predictions? So it would be a good thing if we were to see if we can find ways to distinguish those who may just be trying to make a living (some would call them con artists) and those who may be real (if any exist). There have been many notable people, past and present, who have made it their mission to see if they could find *real* paranormal claims. It's a stellar list of individuals: Harry Houdini, James Randi, Joe Nickell, Ben Radford and many others. The task is not an easy one because those who are trying to trick us are very crafty, and also **if** there are true paranormal phenomena out there, by definition they not something we've encountered, measured, and understood before (i.e. they are not **normal**). But it is striking that as earnest as these investigators are, and as much time and energy they have spent (and money, at least millions, see randi.org), there has not yet, as of this writing, been anything found that can be confirmed.

We have all heard of the many methods of the paranormal such as tarot cards, oiuji boards, rune stones, and on and on. How science treats these is to use the famous "Occam's razor", which says that **if** we can come up with an alternative that is more simple (relies on fewer assumptions), then let's go with that. Occam's razor doesn't **prove** one explanation over another. But it is always a good idea to simplify things as much as possible. So if we have two explanations, it is a good strategy to use the one that is more simple as long as it continues to be consistent with observations. It is not to say that paranormal phenomena are impossible, but if we're going to have a reliable way to predict the world around us, to predict future events, it will be easier on us if we have simple models to follow rather than complex ones. Scientists and researchers alike readily admit that if something comes along where our *equal but simpler* explanation doesn't work, we would have to reconsider the paranormal explanation. Even that approach can get a bit complicated, in that very complex models sometimes are built to explain new information as it is uncovered (epicycles and string theory being some of the most famous). However they are generally still less complex than assuming a new force of nature that no one has ever heard of or detected before. What happens more often is that a new, simpler alternative model will be proposed that still is something in the 'normal' world rather than the paranormal world.

That's not to say that everything is understood. There is one area that I find fascinating that has been given a name, and some possible ideas of how it *might* work, but is far from being understood. Is this evidence of paranormal capability in all of us? I don't think so, but if we're to be objective and honest, it still goes unanswered. Scientists call it the placebo effect (and its counterpart, the *no*cebo effect). I would imagine most people have heard of it. It is, to a large extent, predictable (meaning we can anticipate when & where it's likely to show up), it's repeatable … but it's not very well understood. The placebo effect

33

is when someone's illness, or condition, improves only because he *thinks* he should be getting better. It's most commonly brought on by tricking the patient into thinking they're receiving something that will help cure their ills. Not in every case, but a surprising percentage of the time, people will actually get better; essentially all by themselves. There are some natural explanations, which have to do with the idea that reducing stress allows the body to better heal itself. That sounds good, and perhaps that is the case. Even so, the placebo effect is still, in my opinion, one of the more intriguing areas of science that is still not well understood.

But returning to the vast majority of paranormal claims ... it does seem to be true that either they can be more simply described with normal ('natural', 'materialistic') explanations or that they cannot be reproduced in ways that reduce the likelihood of self delusion, or purposeful delusion (fraud). Then it's quite a puzzle, as I mentioned early on, that fortune-tellers, mediums, diviners, have been around since the dawn of civilization. If they were not **true** then why would they have such great staying power? What are the motivations, besides the obvious of trying to swindle money from the gullible, that spiritualists would have?

Chapter 8 : Why do we pay so much attention, to 'spiritual' claims?

There are a couple of obvious reasons we've already touched upon. The first is the desire to be able to know, and perhaps change, the future when we feel we don't have that control ourselves. It's understandable that some may look to others who claim better skills in predicting the future than the rest of us have. In finance we might call them fund managers. In business we might call them predictive analysts, in medicine we might rely on sophisticated *expert systems*. But in the softer sciences, in the social and psychological world that is messy, it is perhaps understandable that some people turn to those who give them hope so that they might have more order in a chaotic world.

We all do that. No, we don't all go to a palm reader or stare at our tea leaves for guidance, but we do ask our friends for their opinions, and we do read as much as we can about the recommendations of educated experts out there. So looking for others to help us predict the future is not only common, it's definitely recommended unless you're naive enough to think you can master all the complex business, scientific and social subjects out there enabling you to make your own decisions on all topics. So really, as we've said before, it comes down to trying to determine what, and who, you can be confident of and what, and who, you should be cautious of. Appropriately, scientists call this the *confidence level*, and they strive to raise that level as much as they can. But there are other more subtle reasons that can help explain the motivations of both the mediums and those who seek their advice. It has to do with the importance of being the first to know, of having access to privileged knowledge, of hoping to achieve increased social stature and possible material benefit for having access to special information.

Again, it's the strong desire to try to avoid bad things from happening. We all welcome good news, but we definitely pay more attention to bad news, and especially ways to avoid it. It's interesting to look at the *economics* of being a seer.

If a medium predicts a windfall, and it doesn't happen, we may think he is a quack if we took action on his advice. At least there was no big downside, other than perhaps losing some resources since, of course, he most probably asked us to pay for his prediction. But, if he predicts disaster, and we don't take his advice, not only do we suffer if it does happen, he also gains credibility since his prediction of dire consequences came to pass. Conversely, if he predicts something awful and we follow his instructions and nothing bad happens, he still gains credibility since he can claim that following his recommendations was what saved the day! When you work out all the possibilities of predicting good or bad, taking action or not, and whether or not the event turned out as predicted or not, the likelihood of being seen as *possibly legit* is 10 out of 12, or 83% ! It turns out the only real risk in the medium-business, is if you predict something good is going to happen and it doesn't come to pass. (See the Statistics of being a Medium in Appendix 1)

So, I hope you can see that the real emphasis is not spending a lot of time debating whether or not it's possible that paranormal claims are true, but rather to hone our skills on ways to determine which are definitely **not** true. Let's look at some ways to do that.

Chapter 9 : What are the tools & methods for predicting the future?

Models

When it comes to predicting the future, there are really only two options. They're not mutually exclusive, but they're very different. There are of course appeals to the cosmos, contact with the spirits, communing with nature and access to that *special* kind of knowledge as we have discussed previously. Science takes a dim view of these approaches, but only because they can't be accessed through the (currently known) scientific toolbox for study. Knowledge gained in the spiritual fashion is, even admittedly by its practitioners, unpredictable, vague and sometimes ambiguously pernicious. There's not a whole lot more one can say about knowledge (of the present or the future) obtained through these means, even if they do actually exist. That's it. If someone truly does have *the gift*, that would be wonderful. It's just that there are also so many creative thinkers out there too, who would be happy to part this fool from his money, that the best efforts to find *real* spiritualists from the shady, so far, has turned up none who are genuine. So let's move on.

The other method for predicting the future reliably is to assume there are understandable ***cause and effect*** relationships in nature, and then to attempt to somehow uncover these relationships for the purpose of developing tools and methods for detecting and hopefully controlling the causes and so their effects.

This approach relies on another important idea, one that most of

38

us take for granted and that is the assumption that the future will look, to a larger or lesser degree, like the past. By studying the past (i.e. developing a model) and inspecting the current (i.e. collecting relevant, reliable information ... 'data') we might be able to forecast (with varying degrees of confidence) into the future. One of the major ways we try to predict the future is to examine the past, see how the current may be similar to any past-situations and assume the future will have similar characteristics to 'the now-known future of the past'. That bit about confidence is also where the challenge lies. Developing models and fitting the current situation into them is not excruciatingly difficult. As we've mentioned before, our brains do so constantly and we'll have plenty more to say on this very soon. But it's making those models as accurate as possible that is the difficult part.

Model building has been a subject of study for millennia. And there have been great advances in the past decades, with the development of fast, cheap computers. But model building still presents challenges whether it is being done naturally (through our own personal 'neural networks') or through computer programs. One prominent academician who studies these efforts describes the struggle clearly. "All models are wrong, some are useful."[9.1] But others have noted that even this difficulty isn't as dire as we might think. "Even a wrong model is good because it can allow you to make decisions in the right direction."[9.2]

There are three important things about the past; '**what**' happened, '**how**' it happened (i.e. can some cause & effect be established?)

and **'why'** it happened ('why' involves the motivations of people). None of these three, especially the why, is usually without controversy. But even the what is open to interpretation, since learning about the very distant past relies on some pretty sophisticated techniques (e.g. radiometric dating, earth-pole-magnetic-shifts, etc.) and even the not-so-distant past often seems to have some disagreement.

We measure (even if it is a measure of intuitive evaluation, i.e. *gut feelings*) "then" as well as "now", to gather data for analyzing (for generating our own mental-predictive models). We evaluate this data, and further develop/refine our models. But how do we build those models? It is common to use one of two methods. First we seek to analyze historical data and look for factors that seem to correlate. If we hope to establish some cause and effect (so we might be able to change the future to suit our wishes) then correlation must be there, and it is far easier to establish correlation than it is cause & effect. Secondly, if we have little or no historical data that appears useful in a novel situation (including asking input from others), then we are left with no option but to try something. We see what happens, and adjust until we begin to detect patterns and correlations. This method is actually best served by generating data (outcomes) by purposely designing experiments to see how our adjustments affect the results.

But there is always the risk that the future (even in the near term) doesn't quite match up to what has happened in the past. In fact, this is always the case, otherwise we'd have a sure-fire method to predict the future. Christopher Finger of the Risk Metrics Group put it nicely in a Jan'2008 article entitled "The once holy grail"[9.3] "[Mostly] we are well served by historical observations: the past is a good indication of [the] magnitude of future [events]. Surely, plenty of disclaimers apply here, but historical data is what we have, and more often than not, it is of some use. Partly, this is true because [nature is] more or less continuous and thus chances are we will have observed something in the past, and that something will at least look somewhat like what could happen in the future." This study of the past, collecting information/data to build models is part of the large body of science and mathematics called statistical analysis. Ever wonder if study of math and statistics could really be useful? Note the success of one statistician[9.4]: She won a lottery of $5.4M, 10 years later another $2M, two years later still, another $3M and ... just to drive the point home, another $10M in 2010. While she hasn't specifically said what her methods were -- calculated odds of this happening were given as 1 in 18 septillion and that a string of wins such as these are expected to happen only once in a quadrillion years – that she didn't rely on luck. Maybe it's time to hit those books again!

Designed Experiments

There is another interesting way to go about predicting the future, and we briefly mentioned this above. It involves purposely changing the present, and seeing how the 'then future' but now the 'current present' unfolds. Mathematicians have dubbed this method 'designed experiments' (or design **of** experiments, or

DOE). This approach involves first developing a possible idea, or model, and then purposely changing conditions to see how the results change. Then change conditions again and again until we see what parts of our model are really affecting the outcomes we're interested in. Those that have a significant impact **and** are things we can change, we'd keep in our model, and those that are not helpful (or predictive) we would drop. The idea is to wind up with a model that both consists only of those important factors, and (also a critical part) a good understanding of how those factors can be changed to obtain the outcome we want. This is a very powerful tool, and its usefulness can't be understated. Again, our brains do this all the time; perhaps most obviously when we practice some skill. Each time we go through the motions, our brains are trying it slightly differently and seeing how things turn out. Those movements that work out better are strengthened and used again.

Of course all of these methods, while done in nature in our brains, have also been studied and improved using the other big toolboxes out there … science and mathematics. Today computer programs do almost all the heavy number-crunching work for us. We collect information (data) and we tell the computer program what model we'd like to use and have it help us refine it. In fact in many software packages available today there is the option of testing out several models (equations) and reporting which ones seem to help the most.

There have been some excellent advances in how efficient these tools can be; nowadays one can give a computer program hundreds of potential causes and it can pick out the most important factors (if, indeed, some correlative relationship does exist). Once these few important correlates have been found, they can be systematically adjusted (DOE) to determine the strength of their likely cause & effect relationship on the results.

A great benefit that comes along with these methods, because of the way the mathematics is developed, is an idea of the confidence we can expect of the predictions that come from the model. That's something that the spirit world is not only sorely lacking, but sometimes even seems to purposely obscure. I don't think we've ever heard a fortune-teller say "there's a 92% chance you will meet a tall handsome stranger".

These mathematical and computer methods have been under development for many years, and there have been numerous textbooks written on 'collecting quality data', 'model building', 'techniques in DOE', 'assessing risk/benefit predictions' … that we won't spend any more time on them here. Science and business have been using these methods for quite some time, and whole careers have been spent both on developing and improving these methods, as well as applying them in real-world situations. It's a fascinating area of study for someone interested in predicting the future. But for now, I would like to turn our attention to some areas that are **not** quite so easy to predict, even though there have been advances in these areas recently as well.

Improving our models

There are really two major parts to a model. The first is the information we're analyzing (the 'data'), and the second is the form of the model itself. If we have a plot of points, is a straight line the best model that seems to fit the data? Or should it be a curved line, or a sine wave or … maybe just a math equation that fits the data, but is difficult to visualize because it has more than 2 or 3 factors?

Fortunately, there has been a phenomenal amount of work done by scientists and mathematicians to help with this kind of question (is our model correct? Or do we need more/better data?) and luckily we mere-mortals don't have to get into all that detail and complication since they've written software programs that make it pretty easy to apply different models to our data and see if we can make use of them.

One of the more amazing concepts available in some of these software packages is the ability to estimate something called "lack of fit". This is an indication as to whether or not our model could be improved by finding more data/information to use with it. Or perhaps our model itself could benefit from some extra terms, or perhaps even a different model is needed all together. There are many different kinds of models available in these software packages; the classic regression types (fitting to a straight-line, for example, but also non-linear models), decision trees (perhaps the most understandable to non-experts) and quite a few others, including 'neural network models'.

44

We won't go into a detailed description of these computer-neural-networks here (whew! that was close), but they have two very interesting characteristics that are important for what we're talking about in this book. The first is that they were originally developed to resemble what scientists thought the brain does. There are *neurons* simulated in the software, that are interconnected and when one *fires*, it will affect simulated-neurons and perhaps cause them to fire as well. Some are connected to 'inputs' (simulating our senses) and some represent the 'outputs' (something like motor-neurons to muscles). These types of neural-network models have been found to be very useful because they seem to do a good job at forming a usable connection between the inputs and the outputs. The downside to computer neural networks is that we don't have very good ways to know what the form of the model is (again, at least not for us mere-mortals). That means it is difficult to understand how the inputs affect the outputs.

In a classic linear-fit-regression model, once it's optimized to the input data, we can see what *terms* are being important and how strong their effects are on the output; for example we might see that when the temperature outside drops by 5 degrees, our family uses 8% more heating energy. We could have also included the number of windows & doors in our house, and found that there is also a link between the number of windows and the amount of heating energy used. This could lead us to the (sensible) idea that having more energy-efficient/better-insulating windows could save energy; or .. perhaps not! Maybe our energy consumption is currently being affected more by how much we're heating rooms

45

we don't use … or maybe something else all together, like leaving the outside door open when we let the cat out (dang cat anyway!). Or it could be something we haven't thought of at all, so none of the inputs we've chosen to study really seem to account for the monthly change in our energy bill. It's possible that for one of our neighbors such a model would suggest they invest in better insulation, but for our house, perhaps a similar model might suggest most of our energy consumption is through family usage. In our case a family-meeting might be in order to talk about ways to conserve energy.

In a computer-neural-network model it isn't clear how each of the separate input factors affect the result (energy consumption in this example). A neural-network model finds the best model for the data it is fitting, but the model components, these terms, are not easily related to real-world things we might want to change. When a computer neural-network model is developed, it might predict how much heating fuel is used, based on all of the information we fed into it, but we would have very little chance of knowing how to change, or improve, our energy consumption by either adding better windows, or adding more attic insulation, or giving a stern lecture to that dang cat. We lost that connection between the inputs and the outputs. (Not all is lost; there is something called *sensitivity analysis* in which we systematically vary the inputs, similar to the designed experiments mentioned previously, and see how much, if any, they affect the output.)

So, a neural network model can be very good for predicting results when new data comes in, but it's not very good for helping us understand what things affect those results. This is important, because it's one thing (as we've mentioned before) to be able to predict the future, but it's quite another to be able to *change* the future. To change the future, we need to have a good predictive model *and* we need to know what things are involved in the cause & effect of the results, so that we might change those causes, to improve the results. Neural networks help us predict what results to expect, but they're not that useful in helping us understand how to change things to modify those future results.

This brings up a couple of good ideas; first that a good (useful) model has both predictive and adaptive characteristics. We can use such a model to analyze new data to make predictions, and then we can also use it to adjust the world around us -- to change how the future will take place. Given these ideas about models, how then do we build & use models in the world around us, and do our brains do something similarly? Let's start with the beginning; a model is often built (both in computer programs as well as in our brains) by an idea, something that somehow grabs our attention, something that percolates up from the information around us, or from information/data that we have collected and are analyzing. We take that idea, and try to fit what we know about the world to it, and perhaps develop what let's call a *theory*. There are, of course, formal definitions of *theory*, but for the discussion here, let's start with a rather casual definition that is more like an idea, a hunch, and develop that a bit further.

Levels of theory

This concept of trying to organize the information around us, developing perceived patterns into some idea that would help us understand and predict what's going to happen in the future, has no doubt been around since humans gained the cognitive abilities to do this *what if* kind of thinking. But it wasn't until the sixteenth and seventeenth centuries when 'models' began to be formalized in the work of natural philosophers such as Francis Bacon (1561-1626), Isaac Newton (1642-1727), Rene Descartes (1596-1650), Johannes Kepler (1571-1630) and many others. This time period is when many historians of science suggest that the principles of the scientific method were developed. But it took another couple of centuries before things were stated more clearly by Karl Popper (1902-1994) when he proposed his ideas about how science works in his 1934 book, 'The Logic of Scientific Discovery'. In this book Popper focuses on the ideas of objectivity, reproducibility and falsifiability to establish what it means to do good science. Let's look at these ideas a bit more, and perhaps discuss how they can help clarify how an idea develops. Instead of the general idea of a theory, let's rank these thoughts, from the simplest statement, to something more useful. Let's use the idea of different levels of theories, or ideas, and as more of Popper's criteria are used, we might suggest that an idea rises in it's theory-level. For statements or ideas that have no supporting or contradicting evidence, let's label that level of idea as a T1 level theory. Suppose your next door neighbor comes over and claims "Hey, that cat of yours keeps coming over and scaring my dog who then makes a mess all

48

through the house! That's a T1 statement ... nothing to 'back it up'. Then we incorporate Popper's concept of trying to *test* the idea. A test can be supportive, providing results that are consistent with the idea, or they can be challenging, looking for results that would negate the idea. Or a test can be inconsequential, which happens a lot more than scientists would like. With one or more confirming tests completed, that support the idea, it would be raised to the T2 level. Suppose that pesky neighbor points out that every time his dog has made a mess, he has noticed that our outside door was open and our cat was outside. His observation of a correlation with the open door, a cat on the loose, and a messy floor ... is at least some, although perhaps weak, supporting evidence for his T1 theory. Maybe we should consider his idea further since now it's risen to the lofty status of a T2 theory. Most ideas would fall into the T2 level, as generally there is some kind of support offered, and the discussion/debate is usually about how strong this supporting evidence is or how relevant. The stronger kind of test is the *challenging* test, since science can never prove a theory to be true. There can be hundreds of these *consistent* types of test results, but they only serve (usually) to generate consensus that the idea is probably the right one. However, it is the goal of science to propose, and then to carry out these challenging, or falsifying, kinds of tests. Maybe an example would be that we actually watch the cat to see if it is scaring the neighbor's dog. We might observe that actually the dog is getting into a frenzy trying to get to the cat that he sees on the outside, and in the process makes a mess. Far from it being the cat scaring the dog,

49

it's the dog that has behavioral problems. This would be a successful 'challenge test' to the T2 theory that the cat is terrorizing the neighborhood. If there has been at least one challenge to the idea that has **failed in its challenge** (strengthening the possibility that the original idea is correct), the idea now raises to level T3. As more and more challenges fail (i.e. the idea, or theory, holds), the confidence in the theory grows until it reaches the highest level here, T4, where the idea has been accepted as the consensus view. This T4 level is where some scientific ideas have been given the label of **Law**; some examples would be the law of gravity, the law of conservation of mass/energy/symmetry etc. But there are other consensus theories that are more often still described as theory, such as the theory of evolution (although many in the biology community would probably be quite happy with the Law label … it's unclear how many, if any, 'falsify' types of challenges have been made to the theory of evolution). But again it is important to note, as critics often do, that theory does not guarantee *truth*. In fact, most well known scientists have made their reputation by either disproving previous theories (replacing them with other theories) or modifying them by adding refinements that are consistent with the new observations. An example of the latter is relativity, which reduces to Newton's gravity when bodies are at rest.

Note that this theory-level concept inherently uses the notion that an idea/theory can be tested, i.e. that it has some predictability character to it. A useful theory allows us to say 'if I were to do A, then B would happen', so that if we do A and B doesn't happen (it was falsifiable) we've mounted a successful challenge and the idea would have to be dropped or modified. The most important thing to point out, for the discussion here, is that the goal is to have more tools to predict the future that we can confidently rely upon.

How about some silly examples of these different theory levels? Suppose I were to assert, with no supporting claims, that Earth's moon was made of green cheese? That would be a T1 theory; a claim with no support. But, suppose I were to then claim that it has become common, in cosmo-geology research, to label a very specific and unusual configuration of minerals as "green cheese" (note the quotes, which are intended to signify that this term is meant to stand out as 'not the usual usage of the term'). With this supporting/qualifying statement, assuming one is willing to accept it for the purposes of the discussion, then my original idea might be raised to a level T2. Of course, this is tricky, since anyone hearing that the moon is made of green cheese, without the accompanying qualifier that the term is meant in the cosmo-geological sense, would likely draw some wrong conclusions, but such is the frailty of human communications.

You could imagine that some other geologist might disagree, and propose a better analysis of the sample, perhaps using scanning electron microscopy, in an attempt to refute that this particular sample's mineral structure is *not* this unusual arrangement, and therefore *not* "green cheese". The other scientist carries out her experiment only to find that, indeed, the structure is consistent with the original claim, and so the green-cheese consistency of the moon might rise to the T3 level. The theory had been challenged and the challenge failed. This is a somewhat humorous example, but this is the kind of interaction that scientists perform every day in their journals and at conferences; in fact, the idea of the moon being made of cheese was indeed discussed at a science conference in 2011.[9.6] We have all heard of some claim in the news that at first sounds preposterous, but generally after a closer read either becomes a bit more believable, or sometimes even more preposterous! Thus the wheels of science grind on. Does our brain do a similar kind of evaluation when it tests its own models?

How does the brain develop models?

We have previously mentioned the concept of designed experiments, and noted that this is also something our brains do as we practice some new skill. But let's look at the process a bit more closely. Recall also that we talked about how it's our nature to pay more attention to possible negative news than positive ("two in the hand" ideas). So one might claim that our brains are focused upon fear. Actually, I'd modify that just a bit, and suggest that our brains work through FEARR : Focus, Evaluate, Act, Re-evaluate, Record.

When looked at more closely, the way we interact with the world, and the way our brains collect information, store it for future use, and *develop models*, is done through the following steps.

Focus

We are inundated by stimulation from our environment; we perceive it, we filter it, we interpret it. It is so pervasive, that we have to go to amazing lengths to "stop the voice", the internal conversation continually running in our heads. Meditation, Zen, and other *silencing* techniques take years to master, with the primary goal of *stillness*. Perhaps for most of us the closest we come to this, is sleep. We spend a lot of time & energy making sure our sleeping environment is devoid of outside stimulus as much as possible. But at some point, something in our environment **gets our attention** and our brains then begin to process this information, comparing it against all we've learned so far.

I am sometimes amazed to find myself thinking of some far off topic and wonder "How did I start thinking about this?" It's quite an entertaining challenge to see if I can trace back, from subject to subject, the trail of ideas that led to the thoughts I had. When I do so, I have invariably found that the original thought was always sparked by something gaining my attention in my environment. Here is an example. One day, I was remembering the first time I rode a bicycle down the long driveway on a farm in the Midwest where I grew up. I reflected on what an experience it was, the exhilaration of the speed going downhill, the trepidation I felt fearing I might crash, the joy of doing something for the first time that "the big kids" were able to do. But then I caught myself thinking, "Why am I thinking about this?" I began to think back, trying to "connect the dots" of recollections to see if I could trace

53

back to the original thought in my thinking-trail. I realized that this was just the last of a couple of memories I had about that farm. Another recollection was one where I was climbing a thick 30-foot rope up to a large limb in a tree. Another was climbing around in another majestic tree in front of the house, jumping from limb to limb (and we wonder why parents worry). Many good memories were created there. But why was I thinking about that farm? I recalled I had, just previously, been thinking about my sister's farm and how enjoyable it is to visit her and her family there. Why was I thinking about my sister's farm? Prior to that, I recalled I had been thinking about the simple life of living in the country, in the Midwest, were I had grown up. But what made me think of country living? I realized that I had, perhaps only 5-10 minutes earlier, found great humor in a billboard I had seen along the highway depicting a hillbilly sitting on his porch, with overalls and a straw hat with his bare feet up on the railing, with his big toe sticking out of a hole in his sock. It was a very impressive, and humorous, billboard! He looked extremely happy and content and my thought train began. That perhaps is a silly story, but true, and it struck me that I had a trail of 'thought crumbs' that went all the way from something outside, the billboard, to fond memories of learning to ride a bicycle many decades ago. Over the years since, many times I have gone through the same process of thought connections, and always found it to be an enjoyable challenge. So far I have always been able to trace back to something I had seen or heard from my surroundings. This is not to say that we don't sometimes sit and ponder some thought that arose completely

54

internally, but I would suggest that more often or not, these are purposeful thinking sessions where one is trying to deal with or resolve some issue (which itself would be a memory of thoughts that had their origins from things in our surroundings). The vast majority of times we are thinking, reacting, doing ... it is because of something that caught our focus from our environment. Another interesting aspect of this external influence has been described recently. Ever walk into a room, intending to do something, and forget why you went in there? Psychologists[9.5] recently suggested it is because, when you go through a doorway between rooms, the change of environments causes your unconscious thinking processes to change tracks, and so get derailed. While this focusing is the first step in this FEARR model, the importance of our brains interacting with our bodies interacting with our environment, cannot be overstated. The term *embodiment* describes this idea, that 'we' are the whole, of the combination, of our brains and these interactions. Numerous neuroscientist philosophers have given this a lot of attention, and is also discussed in authors' Murphy & Brown's book "Did My Neurons Make Me Do It?" Ok, so now that something has our attention, what do we do next?

Evaluate

Much of our brain activity is spent gathering information from our environment, and comparing it to our stored memories. These two steps together give rise to perception. We all know we perceive our surroundings, but what isn't so obvious is how our brain is constantly "filling in the gaps" of partial patterns it is detecting in our environment. This is another of those things that happen so often and so subtly that we don't give it much thought. Said another way, our brains have certain patterns that have been established through our experiences from early on in the womb (more on this below). New information that comes in is compared to all of our individual existing patterns, and some are selected as being possibly similar. Those existing patterns may have also had other information, from the past, that isn't currently coming in through the senses, so the brain 'fills in' what we're currently perceiving, with this pre-existing information from our memories. Sometimes this can be helpful. You can imagine an early hominid thinking, "let's see, every time I've heard that growl sound, there's been a big yellow cat in the bushes." After fleeing the imagined lion, survivors would swear it had been big and yellow since they'd seen it with their own eyes. When in fact they had only heard the growl and gotten out of there as quickly as they could. Of course, those whose models/patterns weren't quite so developed became lunch and were no longer around to compare notes. But, this *pattern*-filling may not be so helpful. Another time, our friendly hominid thinks, "let's see, every time I've gone to the river to collect water, and seen logs floating by, things have turned out ok. There's a log floating by, should be ok." He then became lunch for a crocodile!

So our brains are not perfect (I predict you didn't need me to tell you that). Still, that's what our brains do. They filter through the information coming in, compare it to existing patterns, and select

56

which pattern is likely to be the one that best matches the current set of circumstances. Our brains are constantly *evaluating* our environment like this. As we have developed these patterns/models throughout our lives, we have also been paying attention to how things have turned out, good or bad. These results are also being stored. As infants, we learn good and bad immediately. We experience instant gratification or perhaps the opposite (maybe a scolding if we've done something wrong). As we grow older, we use the results of these earlier 'instant gratification' experiments to estimate the likely good or bad outcomes of our current decisions. So, as the current pattern is being selected, the brain is also checking out how that pattern, in the past, has worked out. The brain "plays out" these imaginary scenarios in the span of thousandths of a second, and notes previous outcomes. It is important to note that this isn't just a single-step, one-way, type of process, but more of a flow of comparisons. As the brain is taking in sensory data, comparing it with existing memories, it is also blending the two together, building up a *model* of what it proposes is the currently reality. We've mentioned this 'filling in' before. Daniel Schacter has some very insightful discussions of this and other aspects of how the brain uses and misuses stored memories in his book "The Seven Sins of Memory." The brain cycles through this *compare and include* process repeatedly and with each cycle, current information is compared and blended with stored memories, building up the image/model. From all of this, the brain makes a decision.[9.7]

Act

Given the selection of the most likely, and relevant, pattern that seems to relate to good outcomes in the past when this pattern, or one similar, has been present, our brain initiates interactions with our environment meant to achieve the good result it is expecting. The brain and body "take action". There is another alternative, perhaps used more often than action, and that is to do nothing at all (which could be seen as an act itself). It could be because no predominant existing pattern stands out, so we are paralyzed with uncertainty, or perhaps during the time we are trying to evaluate the likely good/bad outcome, the opportunity to act passes ('analysis paralysis' i.e. we've just become lunch.) Or perhaps we've noted that in the past that doing nothing is itself the wisest action to take (that large vicious yellow cat may just pass us by if we don't move.) The end result of our evaluation, arising from the initial 'focus' of something in our environment, is to either 'do' or to 'not do.' And this action/no-action is simply aimed at an attempt to achieve the best outcome our brains have predicted, based on past experiences. It could be (admitted conjecture here) that another of the distinctions between humans and other animals has to do with the restraining aspect of 'no action'. Many animals will freeze in the presence of a predator ... that's not exactly the no-action I'm referring to here. But rather, in the context of how we predict the future better than other animals, humans have developed the ability to purposely do nothing *while* our brains are in the evaluate stage. While the environmental cues are being passed through our existing patterns, and the associations of

58

those patterns with the good or bad outcomes that resulted, our brains make sure that we don't prematurely "take action" while we are doing this filtering. It is possible that this "hold off for now" capability is part of what gives us our imaginations. Perhaps these capabilities are related to our *working memory*. Perhaps the more things one can keep in mind, the easier it is to link those series of thoughts into an imagined-future narrative. Experiments could untangle this; give subjects more things to keep in their working memory and see how successful they are at forming a plan to solve a puzzle that requires several steps. Our previous memories almost *come alive* when our brains are looking through them, searching. It is crucial, of course, that when we are doing the evaluation, trying to sort out which patterns are relevant and which from among those, have resulted in the best outcomes for us in the past, that we don't actually, physically, perform those scenarios until the evaluation is done. In this book I have tried to emphasize the importance of being able to predict the future, and this is perhaps one of the most important steps. This is something that we humans do that other animals do not (or at the very least, do less) that gives us the ability to plan further and further into the future. It is this ability to look at various memory patterns, to recall their outcomes (but not to actually take action yet), and to connect those outcomes to other patterns again and again. resulting in a potential chain of events that can stretch into the next few seconds or into an imagined future years later, that humans have developed better than any other species. We then can only await the outcomes of those analyses and actions, to see how things turn out.

One additional note about these two steps, the 'evaluate and act' stages. Most of the time these analyses and decisions are happening in the unconscious. A much smaller percentage of the time we *consciously* get involved in this decision process, trying to evaluate the situation, trying to evaluate potential outcomes and trying to decide the best actions. When our brains are trying to decide how far to extend one leg and slightly bend the other (to step down from the curb), all of these processes are taking place both very rapidly, and very not consciously. How and when does the transition happen between when our brains are on auto-pilot, so to speak, and when *we* get involved in the thinking? I briefly mentioned above that it is a bit trickier if the brain expected a good outcome from the chosen actions, and things didn't turn out as well as it planned. In that case, the brain has to record for future use, both good outcomes from distant past experiences, and the not-good outcome from the current experience. If another similar situation were to arise again, both of these memories might be used in the evaluation. These kinds of ambiguities lead to more difficulty deciding, and in the extreme may lead to what we describe as anxiety … which of course none of us ever experience! I suggest that this difficulty of deciding between actions to take results in a delay in the transition between the evaluations and the actual act/no-act stage. The brain is still trying to make up its mind (sorry, pun intended). Note that in these FEARR steps, the process could be very quick (someone tosses you a ball and you reach up to catch it, all of these steps are involved even in that short instant), or it could take much longer while you ponder. In fact Nobel laureate Daniel Kahneman has described two different systems of thinking. He describes them as System 1 and System 2, and that the difference is whether or not your brain just takes care of business on its own or whether you become consciously aware of the deliberation process. Clearly our brains are making predictions and deciding upon actions all the time without our being consciously aware of it. This is Kahneman's System 1. When we become aware, and seemingly participate in the decision process, he labels this as System 2. These two systems describe the distinction between more automatic actions and reactions and more thoughtful planning. Point of Inquiry podcast (12/5/11) guest Robert McCauley mentions Kahneman's research and suggests that these systems are closely related to what we generally think of as the difference between intuition and reasoned thinking. I suggest that the difference is also related to the confidence level that the brain has over its proposed actions. The more confident the brain is in the patterns it is selecting, the more 'automatic/intuitive' our behavior becomes. So much so that the process doesn't rise to the level of consciousness. But when the brain's analysis results in ambiguity and uncertainty, it takes longer, hesitation arises and the more thoughtful decision-making/analysis systems are brought into the process. 'I' start thinking about it. This is also closely related to something we are all familiar with, but none of us really understand ... emotions. When our brain is humming along, perceiving, evaluating, deciding, and the near future is happening like our brains expect, all is well. When something in our environment causes our brains

61

to access memories that perhaps conflict in their good and bad outcomes, not only does the decision process take longer, and *higher systems* are brought into play, but other unconscious actions are also invoked, preparing for possible negative outcomes. This is anxiety. The brain and body's very quick response of "fight of flight" kicks in. This preparation has to be quick, as while we are spending valuable seconds deciding, that lion might pounce. Even in today's advanced technological civilization, our brains and bodies go through this scenario many times a day. Worse than anxiety (which is bad enough), there are times when our brains aren't having a tough time deciding between actions with good and bad comes, but rather can only predict bad outcomes and have no good actions to propose. We transition into the other strong emotion we all dread ... fear. These "emotions" are not some mysterious *other self* that is running around in our brains, dueling with our logical selves, but rather just the same FEARR process that runs into ambiguity, or worse, into that situation in which we can only see bad outcomes. Back in the days of trotting across the savannah, our brains and bodies developed the ability to very quickly respond to these dangerous situations, and those who got better at it survived more often. We call these feelings something different, emotions, because it's not something that we are consciously doing, and in fact it is very difficult to have conscious control over one's emotions. What we can **learn** to do is to try to control our actions when our emotions arise, but none of us can stop our emotions without *retraining* our brains, essentially giving them a newly developed pattern with a good result that can be retrieved when similar situations arise in the future.[9.8]

Re-Evaluate

Once our brains have gone through this process of perception, evaluation and 'deciding upon an action' then again, as mentioned previously, we also take note (i.e. perceive and then strengthen the neural circuitries involved) of how things turned out. Then we can use this new information (which will have become a memory) should this situation ever arise in the future (or one similar). This re-evaluate step is crucial, since without it we wouldn't have had any previous outcomes (good or bad) to compare against. Outcomes can either be determined to be good, in which case the connections between 'the actions I just took' and 'here is the good outcome that happened' are strengthened. This is also true if the action taken was to avoid previous outcomes; either to take alternative actions from those chosen in the past, or to do nothing at all. These alternatives, if they avoid a bad outcome would be a 'good thing' (avoidance of bad) and that would need to be recorded as well.

Record

I've mentioned how the brain refers to previous experiences to both select patterns, and evaluate previous outcomes. But how is this accomplished in the squishy cellular biochemistry that's buzzing around in our skulls? You've all probably heard how brain cells – neurons – work. There are sensors throughout our body that detect our environment (light in the eyes, vibrations in the ears, heat and pressure in the skin etc). Once something has affected these sensors, a nerve cell *fires*. Aside from some quick-response, reflex, connections (the 'knee jerk' reaction), most of the time these information signals make their way to the brain. It's mind boggling (pun intended) to recall that there are many thousands of these signals coming into our brains constantly. If nothing else just think of the number of points of color our eyes are picking up from whatever we're looking at in each brief instant of time. When those signals do come into the brain, they will be connected to thousands of other nerve cells. There are estimates that the billions of nerve cells in the brain make trillions of connections with each other. Mind-boggling indeed.

Maybe the only information that one single nerve cell relays is "there's a tiny point of blue just over there in the visual field". But, of course, there are thousands of other cells also noting that there are other points of blue nearby, and perhaps close to those points there is some red, and so on and so on. In one simplistic way, that's all our brains do. Nerve cells fire and relay their single bit of data to other nerve cells. But that's where brains really do the heavy lifting. These thousands of pieces of information are combined. Whether or not brain cells in this flow of information fire or not is dependent on the strength of the connection between the cell that is firing now, and the cell it is 'touching' (the synapse). It's that strength, whether the single signal will cause the next nerve cell to fire or not ... that is memory.[9.9] It may be that, by itself, that single input neuron wouldn't cause the next one to fire; but perhaps if there are two, or three, or more neurons

64

that are also 'coming in' to this next nerve cell, only then will it fire. In doing so, it then signals its activity to other neurons it contacts through its synapses. There are thousands of nerve cells that may be firing, indicating that there is something of the color blue in our visual field. Other neurons may fire when this 'blue thing' is detected to be moving across our visual field. Still others signify nearby colors, shadows, sounds, etc. This then, is a pattern, a perception. We can think of this in a way that makes sense to us. "If the light is red, and if the car I'm in is moving, and if I'm the one driving the car, **then** I'll put my foot on the brake." Notice the entire FEARR pattern: the perception, or **Focus**, upon the red light; the **Evaluation** of the situation (all 3 conditions being true) and the recalling (pattern matching) of previously good outcomes if I were to apply the brake; the **Action** decided upon, of applying the brake, and the presumed good outcome, and the **Re-evaluation** of stopping at the intersection. Then the actual good outcome (happy I didn't have a collision) strengthens, **Records**, these connections so that the next time I need to stop at a red light, I'll do it again.

But don't forget all of this is happening extremely quickly. Realize also that the individual points of information are extremely minute. There are no brain cells that equate to "red light at the intersection", but rather just "a point of red" somewhere in our visual field. Thousands upon thousands of nerves are involved for us to recognize "intersection", and "movement" and "car" and the touch of my hands on the steering wheel, and the feedback sensations as I lift my leg, and the pressure of my foot on the brake pedal. So it is this symphony of tiny   bits   of   information,

65

whirling around in our gray matter, being assessed and compared and strengthened … that is how our brains work. It's all about the patterns we accumulate, and strengthening the ones that work out well for us … i.e. memory.

Note also that all of this is being affected at the atomic level, at the level of Brownian motion, of random movements of chemicals … all of which means that it is impossible to have **identical** conditions, that would result in our making the exact same decisions the second time around *if* those conditions were ever the same. (This is the premise of a claim of those who assert we do not have *free will*. See Appendix 2, and an excellent article by Christof Koch in the May/June issue of Scientific American Mind.) As for the impact of all of this on our use of memories, perhaps Professor Michael Gazzaniga may have summed it up best, (in his book 'The Ethical Brain') "Memory is not so much a mechanism for remembering the past as much as a means for preparing us for the future."

Chapter 10 : Intelligence

When our brains are using these stored memories, comparing different patterns, mixing in other loosely related patterns (something we call 'using our imagination'), and assessing possible outcomes to select one to use ... do we have a term for all of that? I would suggest we do; I think we call that **intelligence**. As I mentioned in the introduction, there have been numerous measurements of intelligence that have been developed over the past century or so. There are tests of delayed gratification, tests of verbal skills, spatial skills and reasoning skills ... there are a dizzying number of abilities that intelligence tests have been developed to evaluate.[9,10] I would suggest that all of these are looking at one or more of the abilities we've discussed here: the ability to record and later retrieve information (memories); the ability to sort through those memories and compare them to sensory data, selecting the most relevant memory patterns to consider; and the ability of projecting possible actions/non-actions onto possible future outcomes to achieve a favorable goal. Some criticisms of intelligence tests indicate a bias to a particular culture or type of education. These tests perhaps would put too much value on just the ability to retrieve a particular selection of stored memories (i.e. facts that have been taught). Education is primarily aimed at increasing our stored memories, but here are also large doses of exercises directed towards improving problem solving skills. All in all, the more one is able to retrieve relevant memories, and *imagine* how future scenarios would play out to

67

one's advantage, I would suggest we would consider individuals who possess superior abilities to do so as 'more intelligent.'

So what have we humans done with this remarkable predictive machine, our brain? How do we use these wondrous abilities to make the world a better place, to ensure a safer future? As with many things human, we have created a lot of good and we have also created some bad, but the good news is that there is still a lot of mystery out there to wonder about. Let's look at the two most prevalent methods we humans have developed, for peering into the future.

Chapter 11 : Science & Religion

Of course the two most common methods we humans have used for predicting the future are science and religious faith. To start with, I think I'd like to propose a distinction between 'religion' and 'faith'. I'd like to suggest that religion be considered 'applied faith', just as engineering is often described as applied physics. The reason this is important, is that there are a whole lot of religions out there. But the idea of 'faith in your beliefs' is a common theme that seems to be universal. If you happen to belong to one of the monotheistic religions (as opposed to some multi-deity religions such as Hinduism, or some non-theistic systems such as Buddhism or Taoism), then your faith would be in the belief of your one deity or the teachings of founders such as Siddhartha Gautam. It is also an important distinction because the conversations and debates that are out there, and that consume so much time, energy and money, tend to be around religious *claims*. Virtually no one claims to be able to "prove" that your faith is wrong. It is essentially the definition of faith that a belief has (and needs) no earthly, or materialistic, justification or proof. You certainly **can** have some reasons for believing what you believe, but it's not necessary.

So, let's dispense with any debate here on that familiar battleground. There are plenty of combatants on that field already, and I would just as soon avoid injury. If you have faith that there is an entity out there that has some influence on your ultimate universal existence on some ethereal plane ... excellent. No one

69

can prove you wrong. But, the instant you begin to define that entity, or its powers, or any of its characteristics, you invite scrutiny and critique. This is both for the reason that you are bold enough to make those claims in the first place (unless you've been one of those lucky few who have had direct, personal, unambiguous communication with your deity) and also because those statements can then be questioned and become targets of debate and possible attack. It's best to clarify your faith in your own mind, decide clearly on what you as an individual can and should do, with, for and/or through your deity and then do it. But unless it's very clear to you that part of your mandate is to go out and actively spread your faith by crossing that line (since you'd have to, to communicate anything about your faith to others), it seems to me to be better to just live your life well, as an example for others to follow. Having said all of that, if somehow your entity imparts to you some aspects of the future, then there's nothing much the rest of us can say about that. Recently, however, neuroscience has begun to look at certain aspects of brain function that seem both to impart that spiritual feeling in our brains, as well as sometimes the impression that we may be hearing voices, real or imagined or transcendent.

So let's turn our attention to the other method out there for predicting the future: Science. In the fairness of equal time, perhaps we should try to define science as well. But that's equally as difficult as is defining religion. I might suggest that science is what is happening when someone is using the scientific method. That sounds good, but it's also surprising that there isn't a universal (dare I say cosmic?) definition of the scientific method either! However it appears that most philosophers and historians of science give a lot of weight to the ideas of Karl Popper. Popper focused on a couple of concepts that still seem to grasp the main idea (although there are those that disagree with some implications.) Popper suggested that the scientific method really focuses on three main efforts.

The first concept is the idea of objectivity.[11.1] It is well known that we humans have weaknesses when it comes to interpreting the world around us. This has to do with the two onerous problems we have discussed previously. One of having patterns that are built up from our life's experiences so that we each have different models, and they're all imperfect. The other is how we each tend to 'fill in' gaps or holes in those models unconsciously so that we don't even know if we really perceive the complete/actual reality of what's happening around us. An important aspect of the scientific method is to develop ways to remove this human trait of subjectivity. This is what Richard Feynman was referring to when he quipped his definition of science[11.2]: "Science is what we do to keep from lying to ourselves."

Popper also emphasized the related concept of reproducibility. If you can describe steps to take that avoid personal biases, and if someone else who has some reasonable amount of relevant technical skill were to follow those steps, then he or she should be able to get the same result that you achieved. There have been numerous examples (e.g. cold fusion and homeopathy, speedy neutrinos) where initial reports could not be repeated, and eventually those ideas lost favor.

The last, and perhaps the most important aspect of Popper's recommendations, is that of falsifiability. This refers to the idea that it is not easy, perhaps it is impossible, to *prove* an idea is correct, but it is much easier to prove it is wrong. If I say I'm the smartest guy in the room, I would be hard pressed to come up with some proof of the truth of that statement that you would accept (especially if you knew me). But it would be relatively easy (I predict) for someone else to be smarter than me on a variety of subjects, thus disproving my claim. It's this idea of falsifiability that is really central to much of what the scientific community does. A claim (theory, hypothesis) is put out there, often with some supporting information that is consistent with that claim. This is not *proving* that the claim is true, but rather just showing why the author thinks it is so. The challenge then has been placed, and anyone who can come up with his or her own counter-claim, again backed by their own solid information, is then often given even more credit than the original author. It isn't necessary that this new information offers a new, better idea (although that would be great), it just has to show why the original idea is wrong. Science marches onward.

Of course there have been whole philosophies and methodologies built up dealing with each of these concepts of Popper. Among them are specifics on what accuracy means. Accuracy involves measuring how far away you are from the correct answer, assuming one actually knows the correct answer. Another method is *precision*, which indicates how far apart the measurements are, even if they're not accurate. Ideally, scientists try to have the most accurate, or most probable, information at the highest level of precision possible. Oh, wait a second, did I mention something important there? I subtly included, "the most probable." In science, as compared to faith, one rarely claims to know the true, correct answer. There is always the possibility of error and/or misinterpretation. And as any schoolboy knows[11.3] the ultimate scientific descriptions of the material world we use are **all** based in probabilities (quantum mechanics, statistical thermodynamics and string theory being examples).

So how does science deal with all this ambiguity? Science does so directly and up front, as it turns out. Science revels in the unknown, relishes the back and forth of data, methods and analyses. Many full encyclopedia size series of texts could be (and have been) written on techniques, pitfalls, unintended errors and even fraud. But by and large, we humans currently know of no better method to learn about the material world, than that of science. After the material world, what comes beyond life and those things that we **can** measure? Well, there, you'd better stick to your faith.

But as for science, it's all about the objectivity of numbers and measurements ... of charts and graphs and plots and curves. What are they all for? The whole purpose (as you'd guess from the focus of this book) is to have a model that allows us to predict what would happen *next* if, given the current circumstances, we were to do A or to do B. It's about not only being able to predict the future, but ideally to change it. But as science generates more and more confidence in models and theories, it all comes at a price. For as I said earlier, in virtually every area of science, the more you know, the more you are likely to get to the point of admitting that it all comes down to some kind of probabilistic prediction. There is very little in science that can be described as certain. There is talk about boundary conditions (beyond which, confidence falls off) and error bars (which are, by far, under-utilized in scientific papers) and confidence intervals (which are like error bars, but tend to change, increase generally, the further into the future we try to predict our answers).

But "how" does science actually work? It actually works the same way that we humans process the world every day. However science perhaps does so more carefully and methodically, using numbers instead of neurons. Our reasoning about the world around us, and the way that science tries to make sense of it, all begins with simple observations. We have discussed how this happens in the brain and how that has profound and subtle implications for how we build our models, filter information through them, and come up with actions. Remember the discussion about how information comes into the brain, gets passed through patterns, one of which is eventually selected (based on the analysis step in FEARR) and then actions are taken, reanalyzed and recorded in the neurons for future use. During this process, there must be some connections in the brain that, when several models/patterns are being compared, signal that one is more familiar and/or has a preferable outcome in which the brain becomes more confident. This familiarity, this *recognition* and various other connections throughout the brain, is at the root of both imagination, and the 'ah ha' reaction (the 'idea' light-bulb that goes off over our heads when we finally 'get it'.)

Recognition of familiar patterns, and the ability to 'put them together' with other thoughts/patterns is how the brain begins the *what if* process that is the hallmark of what makes humans distinct from all other living things. Playing this *what if* game in our head, and testing alternative outcomes, to select one over the others, build upon it, extend it in a variety of ways, is what predicting the future is really all about. And humans do **this** part better and more extensively than any other living thing. This is the process that scientists (and detective authors) call induction. It's taking observations, generating a composite pattern, and coming up with an idea of what's really going on. Academics call this developing a hypothesis. Someone who is really intent on knowing what is **really** going on will then try to test whether or not that idea, or theory, is actually correct. Since, as we've already noted, you cannot prove an idea is true, the best you can do is to try to find ways to disprove it. Here is a scenario that happened to me a couple of years ago that I think might show this process.

We were staying with relatives on a short holiday visit in a beautiful New England home that had electric candles in the windows – they look great at night from the road. These particular candles also have a photo-sensor in them, so that they can be kept plugged in all the time, and only shine at night. As you prepare to go to sleep, the thing to do is to simply unscrew the bulbs so they don't keep you awake.

One evening near sunset, I went from window to window to make sure the candle bulbs were tightened so that they would come on as it got dark. As I approached one candle and reached out, the blub light came on! I was surprised (*F*ocus, something caught my attention.) I thought (filtering through patterns) it was quite a coincidence (*E*valuate, unusual based on past experiences) that the photo-sensor would trigger right at the instant I was about to check on it. So as I began to turn to check on the next window, the light went off! You can imagine my eyes widening, eyebrows rising. Now the focus, evaluate and analyze networks were really going to town. Wondering if there was a short in the wiring or something, I began to approach the light again, and it turned on again! I stepped back (*A*ct, decisions had been made to *do something*) and it turned off. I stepped forward and it turned back on. I was only a very short distance away. As far as I could tell, there was no way I was affecting the light myself, but there was clearly some cause and effect going on. It is a bit sobering to think, that even though this is a trivial (but true) example, it grabbed my attention strongly. These are the kinds of things that might have caused our ancient ancestors to begin to believe in some *otherworldly* connection (except for the electric-light part). But I guess I was a bit more skeptical; I thought to myself (*R*e-evaluate), "There has to be some way that I am interacting with that candle."

I began to inspect the area more closely, without moving (gathering more information, refining my mental model); nothing seemed unusual as far as I could see. I thought I would perform a test (planning, what-if), and approach the candle from a different direction. There was no reason for doing this (imagination at work), but I was just looking for more information. Just as I began to move ... the light didn't change ... (were you expecting it to?) ... but instead, what I **did** notice was that I was casting a very subtle shadow onto the candle, caused by the single overhead light in the center of the ceiling of the room (the *ah-ha* reaction). I stopped again, and slowly approached the candle, making note of where my shadow was falling, and sure enough, as soon as my shadow (which was so faint that I didn't even notice it the first 2-3 times) crossed over the base of the candle (which is where I guess the photo-sensor was), the light came on. It turned off again when I backed away and my shadow left the candle base (**R**ecord the results of the whole experience.) Of course, I found I could walk up and touch the light without it turning on, if I only made sure to keep my shadow from falling onto the candle. This entire episode took maybe 30 or 40 seconds and I had a good chuckle over my childlike silliness. This whole experience made a pretty strong impression on me. For a minute or two there I was really wondering what was going on.

When scientists go through these steps, the ideas they come up with (inductive thought) are often called theories. But, unfortunately, even this concept is so loosely defined that detractors of some theories claim that they're not ***proven*** because "they're just a theory". That is also another interesting difference between science and religion. Earlier we noted that good science would give you a confidence level on any predictions that are being made, while generally religions often profess certainty in their pronouncements. This *certainty* tends to lead to immutability, since if you're certain of something, you can't very well turn around and change your viewpoint on that subject tomorrow. Whereas science is not only *willing* to change it's collective mind, the greatest scientific heroes are those who have done just that; upset the status-quo apple-cart by coming up with a 'better theory'. Science doesn't ever claim absolute truth … only better and better models describing nature.

So, to summarize …
A good scientific result is one that can be reproduced by someone else, and is obtained in such a way that personal biases are absent or minimized. A good theory is one that explains all the known observations and empowers us to be able to predict future outcomes based on past results and current observations. This makes it sound like science is the ultimate predictive method doesn't it? Well, as it turns out, the old saying is true. It goes something like (wildly paraphrasing here): "The more you know, the more you realize just how much you don't know."

Few would disagree with the statement that science has made phenomenal advances. Science and engineering have combined to enable investigators to peer to the edge of the observable universe … both at the large end (out in space) and at the small end (at the atomic and subatomic levels). One of the startling discoveries that science has made, relating to our ability to predict the future, is that we don't have the ability to predict *the* future. The universe is probabilistic. All of you good students out there will know I am talking about our description of nature at the atomic scale, quantum mechanics (QM). One of the pillars of QM is that we interact with just those tiny particles that we're trying to measure, as we're trying to measure them. Werner Heisenberg made his fame by putting forward his theory about his famous uncertainty principle. Regardless of its mathematical structure, what Heisenberg's theory says is that we can't predict anything but probabilities about the characteristics of atoms. Technically, what he says is that the more accurately we measure one component of atomic particles, measurements of other components become less certain.

There were some scientists (some fellow named Einstein and others) who thought that the problem wasn't that nature was probabilistic, but rather that we hadn't yet discovered techniques that allowed us to measure/interact with quantum particles without disturbing them. Their idea was described as the *hidden variable theory*. They proposed that there were some completely deterministic variables (meaning mathematical equations) that would exactly and completely describe electrons buzzing around nuclei.

However, another famous scientist, John Bell, who challenged Einstein's idea about these hidden, or local variables, crushed even this idea. Bell was able to prove that it was impossible. Hidden variables don't exist. Atomic reality is truly probabilistic. It's not important here to get into these details, but rather to point out how the scientific method works.

Someone has an idea (theory) that explains some observable results. Others confirm the results (repeatable) seeking additional observations to see if they're consistent with the original idea or not. Still others try to come up with tests that might prove the theory is inadequate or wrong. If anyone makes observations that are inconsistent with the current idea, then there is a flurry of excitement and activity. Either the original theory has to be adjusted or completely replaced by an alternate, or new, explanation (theory).

These characteristics seem to be the main differentiators between "science" and things that scientists refer as paranormal/pseudo-science/woo-woo/etc (in which some would include Faith). It should be noted that part of what makes something paranormal is its lack of reproducibility. Another common aspect of paranormal is the *experiential* nature of the phenomenon, that it's *subjective*. However, let's be clear. This does not mean that those phenomena do not exist, that they are not true, but rather that the methods and techniques of science cannot be reliably used to study them. One of the repercussions of this is that non-science methods of predicting the future do not have the benefit of reliability and confidence that we get from scientific predictions.

But also note, that science has its own beliefs, since there are things that scientists, and science, definitely don't claim to know with absolute certainty. For example, quantum mechanics simply states as an assumption (not exactly even an observation) that two electrons cannot occupy the same quantum state. This is why you can't press your finger through a tabletop. No one knows why this is so. Scientists take it as a matter of faith (scientists prefer to call them assumptions or axioms of their theories).

This then brings us to an interesting philosophical question.

What is it to "know" or "understand" something?
In one sense, to know something is to either have a direct experience of it, or indirectly, through the experiences of others. But to *understand* something, I would suggest, is subtly different. To understand something seems to imply having some knowledge of its inner workings, or the causes behind it. It seems to imply knowing something about other generally more simple things that are the cause of the phenomena we're directly observing. We can 'understand' the motion of a tossed ball because we can describe it through the effects of gravity, the speed, and direction it was tossed. This all ties into the ideas we've been discussing about having models, whether in a computer or in our brains, of our experiences. We can say we understand how a light switch will turn on the light, and some of us might even understand how the laser light helps generate the music in a CD player. Using these underlying causes, and the models & patterns our brains have stored, we then feel comfortable understanding how the world will

81

change around us. The more we predict, and things turn out reasonably close to what we expected, the more we feel confident in our models, and the more we can rely on them on the future. This is why we refer to some people as being confident. They literally have confidence in their view of the world and their ability to interact with it. And the more confident they are, the more they are likely to have found relationships (understanding) of how to predict and **change** the world around them. You see now why in the beginning of this book I begin with the simple assertion, "Everything we do is about trying to predict the future."

There are some famous physics concepts that we don't understand in this way. We don't understand how/why gravity warps space. We don't understand how/why opposite charges attract. Given this situation, it's surprising there is such a fervor out there debating science and religion. They are both systems of belief. But there are important and significant differences to be sure.[11.4] Science (as discussed above) is a system that is founded on objectivity, reproducibility and falsifiability. Another important aspect of science is the search for, and ultimately the acceptance of, new ideas/theories when new observations don't fit the existing theories. Scientists who either make these new observations, or develop new theories to explain them, become the heroes. Religion (generally) seems to be different, in that there is strong encouragement to follow the teachings provided, and strong discouragement from questioning them. I am not really trying to cast a 'better/worse' vote here, but simply pointing out the differences between how the two systems function. I would suggest, however, that the scientific method is better at learning about the material world, while faith might very well be the best system for preparing for what may come after the material world. So, can these two systems co-exist? Some say no, others say, of course.

Overlapping magisteria

There is so much 'gnashing of teeth' on both sides of the science/religion debate that it's amazing so many people wish to spend so much of their time and energy on it. So why am I spending so much time on it here? These are the two systems that humans have for predicting the future. It's really very simple. The most vocal (and vast numbers of more calm) science advocates state very clearly that science deals with the "natural" world; the world that that can be measured. Pure faith, however, by definition exists where no proof or measurement is needed. Whether or not the subject of one's faith exists is not a subject for science, given it is a question that does not have a measurable approach. Stephen J. Gould described science and religion as[11.5] "non overlapping magisteria" just for the reasons described here. Given my preference for the definitions (religion as 'applied faith') I would suggest that it's really science and **faith** that don't overlap. When those who have faith put forward descriptions or assertions of their beliefs, and descriptions of things that happen in the material world connected to their faith, then I would suggest they are working within a religion (the material world aspects of their faith). Those claims and assertions are then subject to scientific inquiry.

But let's come back briefly to an interesting question I posed above, that I predict you just quickly skimmed over. Why would so many people spend so much time debating the existence of something that cannot be proven? The answer to this question is important, and we will look at it closer later in this book when we discuss the sociological implications of predicting the future. But here's a teaser in the form of a question … Why is it that people have such a strong need to be right?

But just as important is that these kinds of questions and subjects now begin to deal with those areas of human experience that are **less** objective, less reproducible and less predictable. I've got a gut feeling that looking at these areas more closely might yield some interesting ideas.

Intuition

It's important not to forget, or minimize the role of *intuition* as we both experience the world and also try to predict future events. Keep in mind that our brains are constructing models all the time, and constantly revising them, just as scientists modify their models when using software and experimental methods. A big difference is that the brain has been developed over the ages to take in hundreds of thousands of input data constantly. The brain then filters all of that input through hundreds if not thousands of models simultaneously to select which model might be relevant and appropriate at the time. The brain then uses the selected model to choose/decide whether or not to take some action, and finally then does so (or not). All of this happens in the space of a blink of an eye, unless you stop to think about things. Pausing to give more deliberations is itself an action that can have both negative consequences (while you're pondering … the lion pounces) and positive consequences (now that you think about it, maybe it's *not* such a good idea to have that extra hot-dog). But make no mistake, the brain's primary function is to build, test, refine and, most importantly, **use** these mental models as you interact with the real world. It is just as valid an approach as that of any scientist who is using sophisticated computational techniques. It's just that in the computer the causes being analyzed are comparatively easy to measure, and are probably fewer in number (say, in the dozens rather than potentially thousands). It's also probably true that computational approaches tend to be more objective and (hopefully) more reliable (higher confidence levels). The brain not only has to take in the objective information coming through our senses, but at the same time mix in the emotional content. Not to be ignored nor vilified, emotional input adds social context that can be invaluable when dealing with others. This is very possibly a significant reason why computers are currently having such a difficult time *acting* like people, as most approaches specifically **exclude** any aspect of the emotional, social, content. Even so, whether it's our brain trying to make sense out of the world around us, or our attempt to understand the results of a computational optimization of a regression model, it remains that one of the more difficult things to do is to first determine whether or not the model is relevant and reliable. Then even if you do have a good model, it is sometimes difficult to determine what to do with the results.

85

Our brains are taking in information all the time, much more than we could handle if we had to consciously sift through each and every bit before we make a decision. Our brains go through so much of this filtering of existing models, comparing possible outcomes, pulling in some components of other stored patterns (usually unrelated, that's what makes it **imaginative**) it is not surprising that we sometimes get this *feeling.* We tend to call that our *gut feeling* or *intuition.* If we were really serious about improving our abilities, we would keep track of these *feelings*, and their outcomes, and check to see how often our intuition serves us well. It is also important to realize that our brains are doing this for us anyway, whether or not we're making notes. If things tend to work out well for us, then over time we would tend to trust our intuition more (meaning those connections that represent a favorable outcome become strengthened). If not, sadly, we often look to outside sources to explain our ill-fortune, our bad luck. In fact, it's just that our unconscious abilities to match patterns with our current surroundings, and assess which actions would be good ones to take (or to avoid), are a bit more muddled, a bit more tricky to assess … so we just call ourselves unlucky.

It is very interesting to think about where this unconscious decision-making is usually happening, under what circumstances. It's probably not when you're reaching up to catch a ball thrown to you. Not much gut feeling needed there (although some of us have experienced that alarming feeling that we're about to run right into the fence!). It's also probably not prominent when you're trying to decide whether or not it's safe to cross a busy street (much better to look for traffic than rely on your intuition there!). However in all of those cases both our conscious and unconscious decision making systems are at work. I am referring to perceiving a gut feeling and making that somewhat-unconscious feeling part of your decision making. It seems to me, that the most common situation where we begin to give more weight to these gut feelings is when we're dealing with other people. One of my favorite examples is that eerie feeling people get sometimes when they *just know* they're being watched. While there have been controlled experiments that seem to show there is really nothing to this phenomenon (meaning that people really **can't** tell when they're being watched), I would also think that it is entirely possible that there are unconscious perceptions from the environment that could account for these seemingly strange feelings. I can imagine, for example, that when someone begins to stare at your back, they also stop most of their other activities, especially if they're trying to be discrete about it all. While you would not be aware of it, I think it is reasonable to think the brain is picking up this *drop* in ambient noise, and since the activity would be behind you, your brain would be particularly sensitive to unexpected things going on. That would be a great survival trait especially since Mother Nature forgot to install those convenient eyes-in-the-back-of-our-head. Unconscious analyses of the environment would give rise to an intuitive feeling, and then also generate an emotional response (fear, apprehension, anxiety). In reality, it would just be our normal senses operating to help protect us from that infamous cat about to pounce. Having said all of that, I think it's good to recall

87

what I'd said earlier, that people have tried to study this phenomenon in controlled conditions, and have found no such abilities in their results. What isn't clear is if those controlled conditions adequately duplicate the situation that people have found themselves in when they have experienced this gut feeling. One suggestion would be to have the 'behind the back observer' performing some routine task, such as eating or reading a book or chatting on the phone, and then with a cue, focus on the subject's back. If there were ways to control the amount of ambient noise that the observer makes before and during the stare phase, I would predict we might find that the brain is doing it's job quite well in detecting a drop in ambient noise.

With these kinds of things going on inside and outside our heads, we generally think that we are quite well aware of our surroundings, and that we are pretty much in control (some of us more than others). But there are numerous interactions with our environment that we find very difficult to analyze, manage, predict or control. The actions of other people are probably the most difficult things that we try to predict. Trying to figure out where that thrown ball is going to go? You can get very accurate with experience. Trying to figure out whether or not you can trust that shady person looking at you on the subway? You might find yourself relying a lot on those hairs standing up on the back of your neck. It is strange that our brain has found interesting ways like that to communicate to us, it's own body! Even in situations where the best intentions are involved, sometimes we can 'over think' other people's plans. Upon seeing someone who we think is following us, you might find yourself thinking, "I'm sure they're probably thinking I'm going to turn left, so if I want to lose them, I'll turn right." Was it a bad-guy following you in the first place, or was just someone else living in the neighborhood ... and what made us suspicious of him in the first place?

One of the more interesting, and more important, parts of this book is the discussion of how we develop social patterns, and how well we predict the actions of others. It's called social psychology. We'll look into our brain's attempts to predict other people, and the scientific study of those attempts in a bit. First, let's return to that small inner voice we sometimes notice that we call intuition or consciousness.

Chapter 12 : Consciousness

"Mirror, mirror ... who am 'I'?
Will I know 'me', by and by? "

There are few topics that have been discussed in more depth, for more millennia, than the origins of consciousness. Partly this is because there is very little consensus on just exactly what consciousness is even though we each *know* we experience 'it'. Tom Siegfried[12.1] describes it as "the subjective sense of persistent identity." Recalling our previous discussion that predicting the future is one of the most important (if not **the** most important) things that we do, it is no surprise that discussions of and investigations into the source of our decision making skills have been taking place for so long. We make those conscious and unconscious decisions by the use of the models (memories, experiences) we've talked about previously. We *follow our intuition* when we don't seem to be able to consciously know why we prefer certain actions to others at times.

Scientists and philosophers refer to consciousness as "experiential" meaning it's something we experience, rather than something we "understand". We all know the little "I" that is inside our heads. There have even been untold hours debating whether there is a separate mind from the brain, just because it seems so clear to each of us that *we* must be something more than just chemical signals buzzing around in an ovoid bone structure. Some take the position that there is *only* the brain, and that the mind is an emergent property or that the mind is the brain *embodied* in the world around us (see Antonio DiMassio's book "Embodiment"). But while

90

those descriptions are very useful, they are still just that, descriptions, not real explanations. Others have given up the challenge all together. David Gross, a Nobel laureate has wondered if science will ever understand the origins of consciousness.[12.2]

There does seem to be some general consensus that an aspect of consciousness is related to the sense of "self-awareness" and is unique to humans. Some view even this with caution however, in that there are numerous experiments with animals that would seem to indicate that they are self-aware. Apes, dolphins, elephants and some birds can recognize their own, individual, reflections as being "themselves". This is done by placing a mark somewhere on the animal (either with a pen, or tape) in a way that it wouldn't necessarily know that it had been so marked. The animal is then placed in front of a mirror. Most animals are curious and treat the reflection as another animal, even to the point of being aggressive and attacking the glass. But some animals will focus attention on the mark, often inspecting it closely, and in some cases try to rub it off of its body. They will do this even though they cannot see it directly, but only through the reflection in the mirror. Clearly the animal recognizes that the 'other' in the mirror is itself, and that the mark is on its own body.

Here, I will make the case that this self-awareness is not only associated with consciousness, but is in fact the basis of consciousness itself. There is also the concept of **persistent identity** ... we all know that we are still the same person that we were years ago even while we know that *we* have changed a great deal in the passing years. We seem to know this even though scientists tell us that our bodies are being rebuilt, at the molecular level, every couple of weeks. So what persists? Clearly it's the patterns of memories, coupled with that sense of 'I', that sense of self-awareness. Some philosophers discuss the impossibility of stepping in the same river twice, because the water that you stepped in the first time has flowed downstream. So isn't it a different river the second time? We are the accumulation of our recollections of our experiences over time. Said another way, *we* are the combination of our current *self-awareness* and the continuity of our stored memories (patterns/models). This much, however, could also be said of those other species that also exhibit some level of self-awareness. I suggest that what really sets humans apart, again, is our ability to project our interactions with the world, into the future much further than other animals can. Does this mean those animals that have *some* self-awareness, and who seem to have at least some abilities to plan for the future, may also have some degree of consciousness? Controversial as it may be, I would suggest it is true. We humans have, for far too long, felt that we are extra-special somehow, that we are in a completely different class than any other animal. And I would tend to agree, but I suggest it is a difference of degree, more than kind. We can forecast into the future much further in time, with greater accuracy than any other animal. That is the main difference. People who interact regularly with apes, dolphins and even birds, will tell you that they believe their animals are conscious, that they think and feel. So once again I suggest that *we* are not so much different than the rest of the universe. Not so much different except in perhaps one striking, important way. Of course, I have noted that important difference many times by now. It is our ability to predict further into the future with greater accuracy than any other animal. But still, there is that strange feeling of 'I' … even *if* other animals have some sense of themselves, where would this come from? How can it be just due to chemicals and cell membranes? Exactly how does the brain give us this characteristic of self-awareness? We have already discussed the ideas of stored memories, and using those patterns as models to project into the future. If we could understand how we get the 'I' in us, then we would have what we need to understand how consciousness comes about.

Scientists have discovered two traits of neural processing that combine to give us consciousness: they are mirror neurons with their associated networks, and something called proprioception. We will discuss each in turn, and then see how the combination gives us the "us" we each know we are. Continuing with the concepts of neural patterns that we each use to navigate the real world, a reasonable question to ask is, "Where do these patterns come from?" It's clear that infants are born with some patterns and pattern recognition skills already in place. Newborns can recognize the speech patterns and cadence of their birth mothers, and even distinguish speaking rhythms between their native language and other foreign languages. If infants are shown images of spiders or snakes, they will cry and try to move away. If infants are placed on a clear plastic floor, that actually has an empty room below it, they will be fine as long as the lights in the room below are turned off. But as soon as the room is lighted, and they can then see that they are at a great height (even though absolutely nothing has changed except their perception of their own height), they will become terrified. Clearly, we are born with brain patterns to detect these environmental clues, and with avoidance skills for seeking a safer surrounding.

There are hard-wired parts of the brain that connect certain patterns (seeing a mother open her eyes wide, or smile) with motor movements of those same body parts. Infants have an innate reaction of smiling when they see another smile. They raise their eyebrows, and open their eyes wider, when they see the 'surprise!' expression. But notice not all movements are so "built in"; mothers will take their baby's arms and help them clap their hands. Soon, the baby will clap when expressing joy. Other more complex behaviors can be built upon these. The mother takes the baby's arms, holds the baby's own hands over their eyes, as she does her own, so he links hiding his eyes for a 'peek-a-boo' play game. From birth, there is the constant building-up of these behaviors, linked with pleasure or punishment, that becomes the repertoire we all obtain. In fact, it's the very absence of "appropriate" responses that indicate some unusual neural makeup in a growing child. The key aspect of this learning process is the combination of innate patterns and existing stored patterns. This learning takes place because the infant has both *mirror neurons* that allow them to instinctively move certain muscles in response to seeing (pattern perception) things in their environment, as well as a feedback mechanism named 'proprioception'. When you move your arm, your body/brain is monitoring every aspect of that movement from both the macro level (did my hand reach the coffee cup like I told it to?) to the micro, cellular, level (do I detect changes in the tension of the muscles in my arm?) This latter aspect, on the cellular level, is clearly something we don't notice, or usually think about until, for example, we get a bruise. Then we're clearly, painfully, aware when that muscle is under tension. But this monitoring is going on all the time, in all body parts, and there is an exquisite symphony of maintenance, repair, defense and growth that is taking place all the time that our conscious brain isn't directing.

It's this feedback system that is prominent when we notice the bizarre, and somewhat unsettling feeling when an arm or a leg "falls asleep". Then it seems as if it's part of someone else when we move it but get no feedback of that movement or feelings from its sensors. Two psychiatrists[12,3] have even described something called 'alien hand syndrome'. This is where surgery to treat epilepsy results in the left hand seemingly "doing whatever it wanted". If the patient were to use his right hand to start unbuttoning his jacket, his left hand might rise up and push the other hand away. The patient states quite emphatically that he is not controlling his left hand, and that it is moving entirely on its own. "I have no control over what it does," is the description he gives. What is important for our discussion here is that, clearly, the patient's brain is still controlling his left arm, but that surgery has severed the communication between those brain parts that are still perceiving the environment, those parts that are deciding what movements are appropriate and … those parts that are involved in the purposeful deliberation of actions. The patient can't purposely control the arm, so it is no longer part of his 'I' feedback systems. On the other hand (no pun intended) psychologists can show that the brain can readily accept something that is clearly not part of our body, as being perceived that it is. There is the somewhat famous 'rubber hand illusion'[12.4] in which a subject will place one hand on top of a table and the other beneath the tabletop. On the table where the hidden hand would normally have been, is a life-like rubber hand. While one experimenter strokes the rubber hand with a feather, someone else stokes the 'real' hand under the table. After a few minutes of this, whatever the experimenters do to the rubber hand is perceived as actually happening to the subject's real hand below the table. MRI scans of subjects' brains even show the same perception responses as it would if the real hand were receiving the attention when it clearly is not. The brain's feedback systems are expecting the sensation, and the feedback loops go into action even though the actual sensory stimulus is not actually coming from the patient's real hand. It's this *extension* capability that some scientists believe gives humans an important ability to treat tools as extended parts of our own bodies. This would have been a major advantage in hunting and combat back on the savannah.

It is amazing to discover that this monitoring/feedback isn't a one-way street. The brain isn't just keeping track of what the body does, the feedback from the body actually changes how the brain assesses the world. We all know that simply making a person smile, or laugh, even if the "heart isn't in it", will actually make his or her spirits rise. But the body can also control our emotions further than that. In an article in Science News (July 31, 2010 p. 8) patients who received botox injections to lessen their "frown lines" took longer to read and comprehend negative sentences (compared to positive sentences) **after** they'd had their injections as compared to before the injections. The brain's inability to get feedback that the patient was frowning made it more difficult to process the negative information. So we react to our environment, we plan actions to achieve favorable goals (or to avoid unfavorable ones) and learn from our successes and mistakes. From an early age, we become adept at predicting our movements in response to, and in control of, our surroundings.

Let's look at this more carefully at the neural level. In response to some outside event, the FEARR sequence kicks in (see previously). When it comes to the 'act' stage, after a decision has been made, the brain has already projected (predicted) the desired outcomes. If that involves using the body's arms, legs etc. to achieve that outcome, then signals are sent, in a highly coordinated (in some of us) fashion so that we get up, walk over and change the channel (ok, today we pick up the changer and push a button). All the while, as the signals are sent to the arms and legs, the proprioception sensors are also sending feedback to the same neural patterns confirming that the muscles that were supposed to contract, have indeed done so. There is also information that

97

the pressure on our back-side has decreased (as we rise), that the pressure on our feet has increased, and that it's time to now push off one leg and extend the other. A prediction has already been made as to how long to contract the thigh muscle on the extending leg, and then to tense various muscles to accept the weight as we transition our movement to that leg. The sensors all throughout our body are constantly confirming the changes of the body itself, as well as its contact with the outside world.

Here is the crucial point. There is also a ***running commentary*** we have in our brains, whether or not we consciously monitor it. We are saying to ourselves "I'll extend my leg now, I'll push off my back leg now, I'll absorb the weight on the front leg now, and I'll repeat with the other leg". Obviously it's thousands of times more detailed than just that. But if you were to either describe to someone else, or try to program such a movement using a robot, you would have to spend a huge amount of time "breaking down" the intricate movements that we all take for granted as we easily stride across the room.

So we can all easily accept that this kind of monitoring goes on. But recent research has actually been able to detect this "planning" stage, well before the person making the movements is actually "aware" of what they were going to do.

A LiveScience.com article[12.5] describes studies by Richard Andersen at CalTech, where they could predict arm movements (pointing to a cursor location on a computer screen) by monitoring planning neurons in the prefrontal cortex. This is the detection of this proprioception organization in the brain.

If we step through this act/feedback loop very carefully, and realize that it is going on thousands of times every second, we can begin to see some very interesting things arise. First, there is some analysis/processing that results in muscle cells being told to relax or contract, then there is feedback that the muscles and bones are responding as planned/instructed. Our senses are all collecting this feedback in the form of pressure changes, light changes, wind-across-our-skin changes etc. and the brain confirms that the expected changes in our environment, in response to the instructions the body was given, are indeed happening. Once this confirmation is obtained, there are signals sent to the pleasure center(s), reinforcing that our actions/movements have once again, resulted in a satisfactory outcome. This is true whether it's as simple as reaching for that ice-cream cone, or when we finally gain the coordination to juggle several bean sacks in the air.

Our brains seek confirmation that our interactions with the world are as we expected them to be, as we **predicted** them to be. It's when we decide to give an instruction to our body, and we perceive that our bodies do exactly as expected (well, almost, much of the time) that we feel we are in control. But what about those times we are just sitting quietly, nothing moving … we still exist, we still have thoughts. And this too, is a property that we have obtained over our learning years, much as we learned to balance and take our first steps. We learned one of the things that truly sets humans apart, in addition to our better abilities to predict the future – and that is language.

Many books have been written about humans and language, and it is indeed an amazing accomplishment our species has developed. But for our purposes here, I want to focus on the 'little inner voice' we all have. When you are sitting quietly … most of the time you are having a quiet running conversation going on in your head. Not to say you 'hear voices', which implies someone/something else is speaking to you, but rather you are just going through some thoughts yourself. Thinking about this, thinking about that. This is so engrained in our thought processes that those who spend considerable time and effort to 'quiet the inner voice' describe some amazing feelings themselves. They claim to lose all sense of 'the self'! When they are "being still," this running conversation seems to open up their awareness, so much so that their sense of 'I' disappears into the background.

I suggest that the subtle, but very real, interplay of this constant decision/action/feedback/confirmation coupled with the inclusion of our own language/inner-voice/running-narrative, gives rise to the sense of 'I'. Some animals have self-awareness, that's demonstrable. Many animals have the ability to communicate, from the calls of birds to the dances of honeybees. There are raging debates as to whether or not some animals have language. And so if it is ever possible to determine if animals have this sense of the running conversation, we might not be the only **conscious** beings in our world.

Clearly the brain is monitoring and adjusting to our environment all the time even when we're not consciously aware of it. Even while we sleep we move our body to adjust to the pressures under us. But we're not "conscious" of those movements. However, should there be an unexpected noise or light, we may very well wake up, sometimes not even sure what awakened us. Our brains are aware of our surroundings while we sleep, even though we still describe it as an unconscious state. It is the combination of planning (predicting) and confirming our actions in pursuit of those plans, in combination with our on-going internal conversation that we recognize as consciousness.

Consciousness is the monitoring and confirmation of the *FEARR* process we covered earlier. As there is *F*ocus on something in the environment (for example the coffee cup in front of me on the table), then an *E*valuation is done (drinking some coffee would be pleasurable). It may not always be obvious to us what the original Focus was, but we begin to become *aware* during the Evaluation stage (but not always, we have all had the experience of being absorbed in a good book while eating or drinking something, only later to look down and see it gone!) It is during Evaluation, when the brain is filtering through numerous stored patterns and comparing them with previous outcomes that we also recall the 'I' during those previous situations. You imagine yourself taking similar actions to those in your memories, and you imagine the results in the current situation. When your brain has compared the options and selected one from the rest, then the *A*ct phase is initiated (even if it is to do nothing). During that Action, monitoring of self and the environment continues and leads to the *R*e-evaluation step. Then comparisons are made both as to whether your physical actions were performed/completed as your brain had planned, as well as whether the planned goals/outcomes were also achieved. Those results, good or bad, are *R*ecorded by the brain. If the desired outcome is achieved, the brain sends signals to the pleasure centers which in turn send out signals to strengthen those patterns/networks that we activated recently. If the desired outcome was not achieved, the brain strengthens those networks that would have inhibited the chosen action. Professor Michael Gazzaniga has noted[12.6] that "Conscious experience [arises from] the positive feedback that is set up when the evaluative system becomes its own evaluator." Gazzaniga refers to 'the interpreter' part(s) of the brain that gives us this feeling of 'I' when we propose, and then carry out acts. "I" get up from the chair and "I" walk over to the counter to get a drink. Warren Brown, co-author of "Did My Neurons Make Me Do It?" describes it like this,[12.7] "… [this] feedback [mechanism], it is an interaction of action in the world and feedback, and evaluation of the action, and then new action and feedback. You have this constantly-going loop. And all organisms, all biology is in this sort of continuous feedback relationship with the environment. And the issue is … how is this circuit being evaluated, and its outcome?" All this, again, takes place sometimes in fractions of a second; and sometimes plays out over years.

This feedback, and assessment of whether or not the outcome is as expected is crucial to our being able to learn and control our

102

world. Researchers have even been able to measure how frustrating it is to only hear one-half of a conversation (say, a friend is talking on the phone to someone else). Your brain is anticipating the other half of the conversation, the half you can't hear, and gets frustrated when it can't compare the actual conversation with its expectations.[12.8]

So we have discussed the mechanisms of memory, pattern searching, predicting and acting on the physical world around us. Scientists, mathematicians, physicists, and engineers have gotten very good at developing equations and machines to help us in that regard. But it has been very difficult to predict the future in a particular area. And that is predicting the behavior and actions of other people.

Social Psychology

Why is there a section on sociology in a book about predicting the future? We've discussed quite a lot already about how we predict future events. But mostly we've talked about how to predict natural objects like a thrown ball, or how our bodies (and brains) interact with the physical world out there (the unexpected stepping off a curb when our brains were predicting a level sidewalk). We've talked about using computers and software to build models, and using those models to forecast probable futures, and how much uncertainty there might be in those predictions. We have discussed how our brains go through much the same process, but instead of stored "data" in a file or a database, our brains use stored patterns of neural connections as models. We have also made the comparison between feeding new data into our computer models, letting them crunch the numbers, and how our brains take sensory input as data, and process it as well, filtering through numerous models to select one and then using that model to project what our brains think the most beneficial future will be. Volumes have been written on how we assess benefit for this decision process. See for example "Against the Gods" by Peter L. Bernstein about risk/benefit analysis.

There is, in fact, some scientific data that suggests that the primary significant difference between humans and all other animals, are a few genes that are involved in the neural development of the frontal cortex. This is the region involved in evaluation and decision making that I've been emphasizing when we've talked about our brains predicting the future. It seems clear, that humans have two main capabilities that far exceed the capabilities of other creatures. One is that of being able to forecast probable futures further and further ahead in time. The other is the *imagination* that can connect other neural patterns with those being considered, to suggest alternate, and perhaps novel, future solutions. While we are impressed at how fast computers can do their calculations, we don't often realize how staggeringly astonishing it is that our brains are taking in huge amounts of sensory data every second, processing it, comparing it to our experience (our stored models of the world out there) and deciding what preferred actions to take (even 'no action'), all in the blink of an eye (literally).

All of this is difficult. Professional computational "model builders" can lament for hours about all of the difficulties in dealing with messy input data, selecting which factors are important, and in what combinations of them to use in their models. They even have lively discussions about which *kind* of model to use. If you really want to see experts get carried away, let them get started on how to interpret the model's predictions and the confidence ranges that come out of the computer software. So predicting the physical, natural world out there is really tough even though our brains are doing it constantly. But attempts to predict the physical world around us pales in comparison with trying to predict the actions (and intentions) of other people.

Without a shred of evidence to back it up, I will suggest that the explosion of neural processing capabilities in humans' development, is intimately connected with the increase of socialization among humans. I suggest this is precisely because as one gains better abilities to predict the future (even of other people around you), others are doing the same, all the while competing with you for advantage. As everyone competes for resources, you will try to predict what I'm about to do as well, and adjust your actions based on what you predict I will likely do. And so on and so on, so that there were great benefits to be had, back in the dawn of human evolution, to be able to predict the world around us (and as noted many animals can do so to varying degrees). The real adaptive benefit was the competition to see who could better predict **each other's** actions.

Perhaps here is a simple example. As quickly as you can, without doing any calculations, just relying on your gut feeling ... which would you say is the larger of these two fractions? 4.3/87 ... or ... 117/82656

Did you pause a bit and do a little mental math, even though you weren't suppose to? bad ... bad ...

If you just relied on your gut impression, I would predict that quite a few of you would say that the fist number would be the larger of the two, **because** the second one has such a large number on the bottom. I would also then predict that a good portion of those who felt this way, would have then gone on to **choose** the second fraction as being the larger, thinking that this is one of those psychological tricks where things aren't what they seem at first. So to 'get the right answer', you selected the one that was opposite of what your gut told you. We've all seen visual examples of this, such as the two lines with arrowheads pointing inward or outward, making the two lines look like they have different lengths when in fact they're the same length   and other similar examples. There are many books on these optical illusions where things are not what our perceptions and interpretations seem to be telling us. But, don't forget, this section of the book isn't about errors in judgment, or faulty perceptions. Rather this section is about you predicting my intentions, and me predicting yours.

If you happened to go through that quick evaluation and tried to guess the "right" answer, even though it wasn't actually what your gut feeling was telling you; then you used this "I wonder what you're thinking" kind of process (by the way, the first number **is** larger, almost 35 times larger). If you had chosen the second one even though your gut told you the first was bigger, then you were doing so **because** you thought I was trying to trick you – point proven?

Physics is the study of the material, natural world around us. Sociology and psychology are the studies of why we do what we do, and how we react and adjust to each other as we are all trying to figure out what all the rest of us are planning to do.

Because it's obvious that this "I know that you know that I know" is happening all the time (even if not consciously) then we can understand how predicting other people is perhaps the hardest thing we could try to do. As I mentioned above, this all most likely became a predicting/developing frenzy way back when humans begin collecting in groups and tribes. It became a bit easier to predict (and control?) the actions of others, the more everyone got to know each other, and of course the obvious benefit was being *in* the group. It was (and still is, but in different ways) dangerous to be out there alone. Being in a group is perhaps one of the strongest psychological needs there is. Safety in numbers, having each others back, cooperation in all things hunting, crafting, defending -- all of these group activities increase the likelihood that the individual, and the group, survive and thrive. This is clear everywhere we look such as on the playground, in our neighborhood, and in the boardroom as much as out on the African savannah. It is true from herd animals escaping predators all the way down to that most important group … our seeking one-on-one relationships. Being in a group of two is not only a main focus of our lives for procreation purposes, but also for the feelings of safety, security and well being that it brings. We spend a huge amount of our brain processing efforts trying to increase our social value in all the groups we belong to. These groups can be formal (paying dues to belong to 'the club') or very informal (just hanging out with a bunch of friends from school). Most of our groups are informal, ill-defined and transitory. Regardless, we all feel a strong urge to 'fit in', but more than that, we all want to be seen as someone who is respected in our group(s), someone who is looked up to, whose opinions matter, who can influence others to our way of thinking and our ideas and plans going forward. We dress so strangers will be impressed (or at least won't criticize), we practice that 'important but casual' look so that others will know that we're in control (but near panic inside with uncertainty as we go into new situations). We sometimes say ridiculous things as our brains are firing a billion-times-a-second trying to come up with the next brilliant idea, or in a frantic attempt to just 'be cool'.

We do all of this to rise a bit in the social standing of the group(s) we're interacting with at the time. Why? So we have more protection, so we can more likely have things done 'our way', so we can **better predict the future that we want**. All of this, as I noted in the introduction, is so that we can increase the likelihood that we, and our closest group, our loved-ones (most often our genetic family, but sometimes just as strongly our chosen 'brothers and sisters') will fare better, be more safe, have more food, shelter, and security.

Interaction with others, as we all know, is not always easy. One of the ways that scientists try to tackle difficult problems such as this, is first to try to see if we can identify the basic needs that underlie our emotions, values and behaviors. Abraham Maslow was one of the first to try to clarify these needs in a 1943 paper "A Theory of Human Motivation", where he noted that, while physiological needs must be met first (air, water, food etc.) they are quickly followed by the needs of safety and security, including of the social kind we've been discussing here. Immediately after those basics were met, the very next thing that Maslow suggested humans need, is the feeling of "belonging", and to feel loved. The two highest 'needs' on Maslow's list of needs are Esteem and Self-Actualization, which he links to morality.

Humans have a strong need to be part of groups, the strongest of which is usually the one-on-one relationship we all seek. But even those who have met their soul mate will turn outward, looking to expand their group-relationships. Animals do so too, and geneticists have developed numerous theories as to how being in a group can be beneficial even beyond the obvious of outrunning your neighbor if being chased by a lion. But there is an even more interesting aspect of group behavior that stymied scientists from the beginning -- that of altruism. This was an idea that puzzled and worried Charles Darwin even as he developed his theory of evolution. In nature, observers would notice that it wasn't uncommon to see one animal sacrifice itself to help its group. This

109

is most notably seen when a mother protects her young. But even this common behavior generates further questions. If a first-time mother were to sacrifice herself for her first-born, and not be around to have further offspring ... would she not be less successful (genetically) than a mother who was to live longer and leave more descendants? Nature has clearly answered this question for us. These are the kinds of questions that scientists like to ponder, to see if they can determine whether one behavior shows a distinct genetic benefit over another. Darwin and others noted that non-related animals would sometimes help another of their species to the detriment of themselves, even sometimes sacrificing their lives. How could this benefit those that gave up their lives if their genes would not continue while the genes of the one that survives would be carried forward into the next generation? If the genes that favor altruistic behavior were not being passed forward, bur rather those genes that were 'sacrificed for', wouldn't the genes for altruism die out? But they haven't. Now scientists had something they could dig into; an excellent puzzle to ponder. How could this be so?

Lee Alan Dugatkin explores this fascinating scientific detective story in his excellent book "The Altruism Equation", tracing science's attempts to put 'other before self' on a more mathematical basis largely through the works of William Hamilton. Oren Harman has gone even deeper, describing the life of one of the giants in the field, George Price who not only developed key insights into how altruism can be passed from generation to generation, but who also took the ideas to heart and gave everything he had to the poor on the streets of London, dying penniless. That, too, seems to strike a cord. If we give and give and give to those less well off than ourselves, then unless we have an unlimited supply of riches, we will become poorer as they become better off. Realizing there are vastly more poor people on the planet than rich (ever wonder why this is so?) ... then the challenge is to determine how much good can be had for the multitudes of the poor, by reducing the rich to the same level. It was a question that George Price apparently didn't consider. He died as poor as any on the street.

So, as with virtually all things in nature, it seems there needs to be a balance. Another famous scientist, Vilfredo Pareto, looked into this question. Around the same time Einstein was publishing the papers that would make him famous in physics (1905), Pareto was publishing his own foundations of microeconomics (1906) by pointing to the large differences in wealth in Italy at the time (20% of Italians owned 80% of the wealth). He developed what is known as the Pareto-distribution, a mathematical description of the ratio of rich to poor that most societies can handle without social stress. The suggestion is that when there are more poor than this ratio, there will be social unrest and political instability.

So that is the rub; how do we do the best for ourselves and our chosen groups (family generally being the strongest and most easily defined) while at the same time preventing such an imbalance that others seek to help themselves to the fruits of our labors? Humans have generally solved this by arming themselves in defense (or taking up arms should one be on the poor side of the equation). Nature seems to have also developed some tools to help in this regard, namely altruism. By helping others, more so the closer they are in your circles of relationships, then the more likely you too will be helped when needed. While sociologists have seen this kind of behavior throughout nature (mostly but not always within the same group/species, such as the social insects), it wasn't until the works of Hamilton and Price that it was shown how this could be an evolutionary, generational, benefit. So the concept of 'enlightened self interest,' the classic mutual scratching of backs, was developed to explain why we help others. It's the reciprocal behavior that is the key. There is a delicate planning (predicting) that goes on all the time – generally unconsciously but also often quite consciously – evaluating who to help, and the likelihood you may receive help from them in the future.

Nature has developed, over epochs of evolutionary development, the idea of being nice. We each practice this and have developed institutional encouragement for doing so. We each have a vested interest in helping others. The more likely we are kind to others, the more likely we will receive kindness in the future. That all sounds nice and comfy and rosy doesn't it? Unfortunately, there is one small twist that continues to make the world a dangerous place.

It's the very same thing that has enabled humans to rise above other animals that can also cause us the most risk. We each have the ability to look into the future, and try to project the results of our actions. Because we are focused on benefiting ourselves and our family, friends and other more or less tightly knit groupings, we compete with others for gain. It's not much of a problem when there are plenty of resources, but of course conflict has always risen, and will continue to rise when there are real, or perceived, shortages. It's sad, but it's true. A recent commentator[12.9] noted (paraphrasing here), "lions and tigers and bears … they're dangerous, but what we really have to watch out for is each other."

When brave soldiers tout the necessity of being "up on that wall" on watch, what they're defending against aren't the savage beasts, but other savage *human* beasts. Even that description is telling. There are two hugely important tasks that military trainers undertake with new recruits. The first is to turn them into a cohesive group, even to replace the family-group back home ("I'll fight for my loved ones back home, but I'll die for my brothers in the trenches"). The other is to dehumanize the enemy. The first thing you must do, to effectively be able to do harm to another is to decrease their humanness, their kinship, in your mind. You must adopt an "it's them or us" attitude; "you're either with us, or against us". And that, surely is the scourge of this wonderful gift of intelligence we possess.

Human behavior is the most difficult thing to predict; if we were able to understand each other completely, no one would doubt we would be able to get along better and avoid the "zero sum" approach to increasing our own well-being, as well as that of our loved ones. But life is hard, and busy, and complicated enough as it is. It is seen as far easier by some, to look out for one's self, and to both guard against others as well as compete and win against others when the opportunity arises, than it is to seek to understand the plight of all and look for solutions that don't result in harm to others.

The general approach of enlightened-self-interest-reciprocity (e.s.i.r) has been studied both sociologically and also computationally. In the 1980's, Robert Axelrod of U.C. Berkeley sought to understand, using game theory, what kind of strategy worked best in a computer game where "rules" competed against each other to see which would do better in direct competition. One could either share points during an encounter, or one could take points but not give any, or any number of other combinations. Many of these 'rule automatons' competed, using a variety of strategies to see what worked best. Over several tournaments, the same simple idea prevailed again and again, called "tit for tat". The idea resonates with the great sages from history. When you first encounter a stranger, be nice, offer a gift, and continue to do so as long as they reciprocate. The result is mutual benefit. But if the other competitor ever seeks to take advantage, then immediately treat them just as harshly in return -- tit-for-tat. Thereafter, offer to 'be nice' again. Even though many smart people tried other, sometimes very complex, sets of rules/behaviors, over many competitions, the tit-for-tat strategy consistently rose to the top of the results. However it didn't appear that the scientists who generated theses strategies tried to imbue those computer entities with abilities to perceive intent, to retain memories of specific previous encounters and then to impose those memories on new strangers that may have some characteristics in common with those who treated them poorly in the past. But we do.

This is at the heart of the question we posed earlier about why people have a need to feel 'right/correct'. Theoretical psychologist Nicholas Humphries[12.10] calls this "Social Intelligence" saying, "It's so familiar to us, we do it of course all the time, we spend all our time wondering what's going to happen next, in our own lives, in our friends' lives ... always analyzing, trying to get ahead of the game, by interpreting behavior, reading minds. And then, like a game of chess, moving several moves ahead ... to see what possible outcomes there could be. Social intelligence became the prime mover in the evolution of primates, including human beings." There have also been suggestions[12.11] that the development of logical thought in humans is directly related to the need to persuade, to convince others in your group that your way is the best way. It is at the heart of confirmation bias. The more you know about what's seen as fundamental (accepted as correct) to your group, the higher you generally will be in your group-standing. It's important to us to be a *leader* of these groups. We strive to have others admire us, look up to us, "follow" us, "like" us (it's no coincidence that these are the two most iconic terms of social media). This comes from the roots of group behavior since the savannah. Not being left behind, not being isolated was surely one of the earliest characteristics ingrained – not only in primitive sapiens but of course many animals also show this important grouping, herd behavior. It's well known by all predators, human and otherwise, that the most vulnerable individual in a pack of prey is the one that stands out from the group. Mechanisms to stay part of the group are so strong, that generally the only way one is isolated is either through injury or sickness. Rarely is it because an animal is excluded or **chooses** to voluntarily separate. So much so that exile or abandonment is seen as one of the strongest punishments against anti-group behavior throughout the animal world. We humans do so and call it prison. And what is one of the worst punishments for bad behavior in a prison? It is solitary confinement. Maslow is right. After we have air to breath and water to drink our social connections, and to be loved and admired, are what we crave.

So, how to best be surrounded by those who will help protect you? It is to have them connect with you, even if it's just by way of admiration. We all want to be seen as, not only **part** of our chosen groups, but also as a good representative member, even as a leader in those groups. It's humorous to watch teen-agers who are sorting out all of these social rules and interactions. It is common for there to be a group, or groups, who distinguish themselves from "everybody else", usually through choices in clothing and/or bodily-modifications (hair, tattoos, various forms of body-painting). But *within* the group, they are individually seeking to be a leader, to be seen as creative, "fashion-forward" (in their chosen style). We each want to be special, important, *within* our groups. What we also would like to do, some of us more than others, sometimes verging on the obsessive or maniacal, is to

116

expand the boundaries of our groupings, desiring to have more and more people admire us, follow us, even if we don't really relate to them very much at all. Some may refer to these types of people as 'politician' or someone seeking fame, or at the extreme, narcissistic. So what seems to be the most significant determinant of group inclusion/exclusion? Certainly one of the major factors is the concept of fairness. Do you treat me as well as I treat you? If not, that's not fair! Time and again throughout our history of societies, one of the main functions of communication has been gossip, relating to each other some news about someone else. This is both for the purpose of establishing unspoken rules of fairness, and also to determine those who do not seem to adhere to the rules (or maybe are just oblivious to them). Eventually those who don't conform to these group *norms* are sidelined and eventually rejected. If you want to be aware of those who pose the greatest threat to you and your loved ones, pay attention to those who feel they have been marginalized, ridiculed, abandoned by the groups they wish to belong to. How well society succeeds in thriving depends on how well society can **predict** who is bordering on the edge of feeling desperation from social exclusion.

So again, what sets us apart (being able to predict future behaviors) can help us achieve safety, or can at least help us in a cautionary vigil. We can spend all of our time plotting how to take from others and how to prevent them from taking from us or we can practice what nature gave us – the ability to see into the future, to do well for others while understanding they might equally do well for us. Sadly, it is true that the real world is filled with a

117

mixture of people who are willing to emphasize one of those paths over the other. The realistic future for all of us is a combination of take, guard, or give. What sets us apart from the animals in the wild is not only the ability to predict further into the future, but the ability to change that future to better suit our goals. Perhaps even more importantly we have the ability to use our memories and patterns, as well as our ability to forecast into the future, to decide *what* those goals could and should be. There has been quite a lot of discussion about the lack of a 'moral compass' available to those who do not belong to an organized religion. But this ability to use our patterns, our predictive skills ... our brains, to realize that overall good can arise from understanding the power of reciprocity, that all humans can *decide* the best actions to take to improve the quality of life for ourselves and those we care about. We can choose to do good deeds, but we must remember that there are other humans out there who, for whatever reason will seek the benefits we have obtained in life. They just may not have adequate mental or social capabilities. They may feel they have been wronged and so are looking to provide (what they feel is) an appropriate 'tit-for-tat' reply. They may feel they are on the lower end of the wealth spectrum and feel it's only fair to obtain their share. It's a long list of justifications that people have when they feel they have been marginalized and socially ostracized.

So what is the best strategy for going forward? It can't be like the computer automaton just roaming around waiting to see how others treat us and responding accordingly. You must realize that what is important is to recognize those enlightened-self-interest-reciprocity opportunities (e.s.i.r.) and use them to lay the groundwork for better relationships.[12,12] This is how you can better predict a more safe and secure future for you and your loved ones.

You must realize that it's not *just* e.s.i.r. but **Your** e.s.i.r. or YESIR, that matters[12,13]. It is how you, as an individual, participate within your groups, and how you and your groups treat those in other groups. This is how you achieve the higher levels of needs that Maslow described. Note that he didn't suggest that these are lofty goals that we all should strive to achieve. Maslow suggested that these are **needs**, and that these needs are what drive us. Once we have enough air, water and food, we seek relationships, and respect ... love and self-confidence. YESIR is the way to achieve these needs ... while at the same time exercising that gray matter to realize that you must be alert and cautious to those who may seek to take advantage. The best approach is to treat others fairly, while being ready to protect yourself and your loved ones. But, realize, that the best approach is **not** to seek to take advantage, because overall, others will see your actions for what they are, remember, and plan accordingly. Those who do not practice YESIR may appear to benefit in the short term, but history and many wise folks will point out that the greatest benefit and security can be achieved by elevating everyone. This concept of YESIR is an extension of the 'maximum utility' theories of Jeremy Bentham and John Stuart Mill (lookup utilitarianism). In YESIR, there is an expanding, layering effect of the groups we choose to interact with, that we consider ourselves to be part of. Closest would be our spouse and children, our immediate family and then the extended family. Most often then those groups are followed by any number of more loosely defined groups such as a neighborhood, or a town, or a club or even fellow fans of a sports team. YESIR says that you are, and should be, more likely to do well towards someone in a closer group, than in a group further out. The essence of YESIR is that each of us defines, both explicitly and implicitly, who is in those groups, which groups take precedence and how quickly the 'closeness' falls off as groups get further and further out from the central point … 'you'.

There is, at some point, recognition of those who are in 'other groups'. In a situation where there are only two options, we are more likely to do well to someone in any of our groups before we do well to someone in another group. But another critical aspect of YESIR is that, when there is no detriment to be had to someone in any of our groups, we would benefit by doing well to those in other groups. This is in anticipation that those others would do the same. Life gets complicated when we add in the reality that there are those that seek to take advantage of others for their own benefit and/or the benefit of their close groups. So, while YESIR has a definite call to do well to others, there is also the need to be cautious. Be fair, but be aware.

Note this concept of fairness throughout. But we haven't really thought much about what it means to be *fair*. Alan Alda, in a 2005 episode of Scientific American Frontiers entitled "Hidden Motives", talks about experiments where he is placed into an MRI scanner, and is offered deals of $1-$10 from another participant.

The other subject (either a person or a computer that is generating random numbers) starts with $10, and can offer any amount to share. If Alan accepts the deal he takes what is offered and the other participant keeps the remainder. But, if Alan were to reject the amount offered, they both get zero.[12,14] If people were purely rational and logical, any amount offered would be accepted since $1 is better than $0. Surprisingly most people will reject an offer of $3, (where the one who offered the deal keeps $7), since they perceive that to be "unfair" and would rather punish the one making the offer even if it means $0 to themselves. Interestingly, however, the researchers found that people are more likely to accept deals even as little as $3 if are told the other participant is a computer generating random numbers rather than another human making the same offer. In either case, they have a choice of receiving $3 or $0. If they are told it is an impersonal computer, they tend to take the $3; but if they are told it's another person, they are more likely to seek a fair deal. This is rarely seen in animal experiments; animals are more likely to take anything that's given. However, recent experiments have also shown that a number of animal types to, indeed, have a sense of fairness. Both chimps and dogs will refuse to continue following commands, if they see another animal receive a treat for their performance when they, themselves, do not. Even in the experiments where animals are willing to take a deal that humans would consider to be unfair, I would suggest it would only be comparable if the human subjects are placed in a similar state of distress where it would be more likely that 'anything is better than nothing'. The animals most commonly are offered bits of food in an environment where food is being controlled or restricted. One wonders how the human experiment would play out if it were someone who is very distressed with a need for food or money. But of course to set up an experiment like that would be considered unethical.

Fairness

Fairness, morals, ethics … these are some weighty subjects. There has been a lot of discussion in recent years as to where these rules seem to come from. Those who have a religious basis easily point to the rules of their faith. Those who don't believe in an existential being, simply tend to state that they can be as ethical as anyone. What is interesting is that YESIR can be the basis of ethics for the non-believers. Once you realize that your favorable treatment of others increases the likelihood of your receiving good treatment later on, then all kinds of rules and guidelines follow that equate to an ethical and moral way of life. The sources and the ultimate purpose of both faith and non-faith belief systems results in the same behaviors.

Given all of this, and realizing the beginning of this section started with the assertion that the hardest things to predict are the actions of other people, have we come to any conclusions that can help? While it's true that YESIR is a good general way to proceed, what about those who roam the country-side seeking to take advantage? Realize we're not just talking about armed-robbers, but also con artists of all stripes, whether it's snake oil or 'water-with-a-memory' as a cure for all ills. Are there tools to predict other people's behaviors? There have been some striking achievements in the recent past along these lines. Most of us are familiar with the history of psychology and psychoanalysis etc, but that's not exactly what I am talking about here. A professor at NYU, Prof. Bruce Bueno de Mesquita has developed a game-theory approach[12,15] that is similar to the methods that highlighted Axelrod's golden-rule philosophy of the tit-for-tat strategy. De Mesquita's methods present a way of "evaluating how people interact when they are trying to advance their own interests."

This approach uses some sensible ways of looking at the people involved in a siutation. It takes into account their goals, their influences on each other, their flexibility, the salience (or relative importance) of each issue being evaluated and a host of other factors. Interestingly, and ironically enough, one of the more significant developments that Professor De Mesquita employed in these simulations was to implement a degree of randomness into the events as they unfold. The professor runs a series of simulations where the individuals-in-the-computer-game make decisions, form associations/alliances and modify their positions on the issues. Over several iterations, and running many simulations, including some of those random events, the Professor comes up with a ranking of likely futures, and how the results affect those who have a stake in the issues. It can be serious business. It isn't just a game. Professor de Mesquita has addressed some of the most important political issues of the times. He has looked at the likelihood of nuclear proliferation. His methods have been put to the test in real-world, real-time situations and his predictions were not only found to be "accurate 90% of the time" but they were also[12,16] "much more detailed than traditional analyses."

What is the key to the professor's success? He focuses on the things we've been talking about here. There are assumptions that key players will act in their own best interests, as well as the interests of their groups/connections. There is a healthy dose of realism in that sometimes people will be cooperative and compromise, but sometimes some people will not. And each of the decisions that the players make, ripples through the other participants, changing the likelihood certain issues will become more popular and potentially changing alliances and so on and so on. When Professor de Mesquita looks at the problem from all of these angles, he has an amazing track record of applying this human nature to predict associations, adjustments, attitudes, and outcomes. Psychologists are making advances all the time in understanding the rational, but just as commonly irrational, choices that individuals make. Researchers like the Professor are also making real strides in understanding how these individual attitudes, biases, and interests interplay to give some light on how group decisions are made. All in all, it seems the future of prediction is looking very interesting.

So where have we been in this discussion? It's amazing to realize that there are a few basic human needs that drive our behaviors, especially when we add in the fact that as soon as we think we know what the future is likely to be, we tend to tinker with it to suit our purposes. But of course, it's exactly this kind of behavior that de Mesquita takes into account. At the end of the day, it looks like the best way to predict the behavior and actions of other people is to get to know them and their motivations. To paraphrase a common saying, "Keep your friends close, and your competitors closer." All the while we should recall that the *tft* strategy suggests that today's enemy can likely be tomorrow's ally. By default, treat others as you would have them treat you, treat those in your groups as they would like to be treated, but always, always, open your consciousness and empathy to try to understand their plight, their stresses and their goals. I predict when you do so, not only will you find that you will gain a better understanding of why they do what they do, but you will also find you will frequently have the same issues, needs and goals as they do. When you get to this point, you will be stepping up to the highest level of Maslow's Needs pyramid towards compassion, and self-actualization. It's a goal we all should seek, and I predict we all need to give it some serious thought.

There is one last subject in this section that deserves some attention. Given that we're all mere mortals, and given the old saying "learn from other's mistakes, you won't live long enough to make them all yourself", there is an important topic that concerns how we choose what information to believe. This subject also involves the concepts of belief and faith, even if we're talking about the most secular scientific concepts. I predict only a small percentage of readers of this book will have studied laser physics, but I would think the vast majority are willing to believe it

125

works just as the physicists and engineers say it works, as we use our CD players and as we scan the bar codes at the supermarket checkout. Even the most highly educated scientists haven't all had the chance to study graduate level quantum mechanics *and* systems biology *and* next-generation genomics *and* clinical psychology *and* <insert your own favorite long list of subjects ...>. Yet, we are all more than happy to believe in most of what we hear or read about these subjects. Even in the sciences, we have learned to accept the pronouncements of experts in other fields with little complaint. However, that doesn't mean it works out all the time; there are dishearteningly frequent stories of academic fraud, and hucksters of pseudo-science and outright quackery abound. So just how is it we decide who to trust when there is a new report? Numerous authors have discussed this subject at length, from Michael Shermer's "Why People Believe Weird Things", to Stephen L. Gibson's "Truth-Driven Thinking", to Stephen Heatherington's "Yes, But How Do You Know?" There is no way I can add much more of significance to their discussions. However, I do want to stress the main point here, and that is **we all** generally learn by gathering information from others and deciding who/what to trust, and what/who to view skeptically.

The reason I bring this up is that I have heard, all too often, scientists and philosophers deride serious people of faith who do exactly the same thing. Recall my earlier clarification between faith and religion (religion being 'applied faith'). When religions make a material-world testable claim, they subject themselves to the lens of science and the scrutiny that comes along with it. But there are many sincere people of faith who have spent considerable amounts of time (sometimes lifetimes) seeking answers to their own 'theories of the universe' kinds of questions. Unfortunately, questions of faith don't have the same kinds of objective tools that are in the scientific toolbox. But they do have centuries of other serious, sincere, travelers who have gone before and have shared their experiences, recommendations, and instructions.

In full disclosure, I am a firm believer in the scientific method, but I do *not* believe that science has the ability to debate the kinds of faith that I have seen arise in others when it comes to their own spirituality. So, science, work hard to test material claims … work hard to question religious claims … but move along when it comes to questions of faith.

Chapter 13 : So exactly what *is* the future of prediction?

There are many serious students of the future, from Professor de Mesquita to 'futurists' like Ray Kurzweil and others, who make it their main goal to try to better predict the future better. What seems clear to this student is that we humans will continue to become better and better at developing these tools. We are obviously living at the dawn of the true *information age* where there is not only an astonishing amount of knowledge and information being generated, but an equally mind-blowing amount of it is being stored, retrieved and analyzed. Recall that there are really two main components to any attempt at future prediction, whether it be computational or a gray-matter-neural-network. Those are the data going into the model, and the model itself, along with the one paramount assumption that the future is likely to resemble the past. We can make better predictions if we focus on the most relevant data to use (while ignoring less relevant data, but then again, sometimes it is not always clear which is which), as well as improving our model that is using that data. We can only marvel at how nature has accomplished this capability millions of years before math and computers were invented. We are constantly "pulling in" data from our environments, and improving the models we use to process that data (we call it learning).

Where science and math seem to have made improvements, in general, to that gray-matter method is by the application of objectivity ... meaning the removal of emotions from the equations. All the while recalling, as we've discussed previously, that emotions are really just unconscious models that are being weighed and offered up to our conscious as possible factors that need attention. However, is the removal of emotion a good thing? Generally yes, but in the absolute, a resounding *No*. Even the best texts on modeling will encourage the student to remember to confirm that computed answers need to match the student's gut feelings.

While computers have become extremely adept at dealing with huge amounts of data, and developing dizzyingly complex models and techniques for evaluating them, computers still pale in comparison with the tasks that our brains do every day. It is ironic that what helps our brains excel in this way, is also where our brains make the most blunders ... emotion. Our brains have not evolved to be 100% accurate because the world is fuzzy. We need to be agile and adaptable as much as we need to be *right* in each new situation. We come back to the concept of filling in partial models, for example only seeing part of the lion's shape in the bushes. If we were to wait until we were 100% confident in the model, we'd probably be dinner. So, we err on the side of caution, and take evasive actions when the partial model seems to best fit the probability that the lion is ready to pounce. We are actually so good at jumping to this kind of conclusion, that our societies have developed strong criteria to help avoid the kinds of mistakes that can arise when we make conclusions based on partial information. We call it the justice system. We have heard that we may allow nine criminals to go free, so as to avoid imprisoning that one innocent person, convicted wrongly. We have a predilection to think we know what really happened, what someone was really thinking, by imposing our own experiences on the situation. If we were always right, we wouldn't need the stringent rules of evidence and law that make it very hard to convict someone if he didn't commit the crime. Still, it happens. The world isn't perfect, we humans are not perfect, and we need the tools of science and mathematics to help us do well in life. But we cannot forget the *human* side of things either such as our relationships, our groups and our spirituality (and I mean this in science as well as in religion). These are also extremely important, and we will do best as we develop the models that help us learn to use the right amount of each to do the right thing, to help others in our widening circles, and to 'give the benefit of the doubt' to strangers. I predict that we will all continue desire to learn and grow as the planet is (socially) shrinking. I predict that while there are those who will try to do us harm, we all can do best when we offer a helping hand when we can, develop the trust we all desire, and share the faith that we can all do better if we just give each other the chance.

Notes & References

Notes and references given here are meant to be more descriptive ("here is where I found this info") than authoritative ("this is evidence for the statement"). Other, more rigorous academic works typically offer their references with the 'authoritative' goal in mind; but here, I am simply pointing to where, and when, I saw or heard (through podcasts) something mentioned, so that others can read further should they wish. I am very interested if anyone has alternative information, and will be happy to make changes as I learn more from anyone if they wish to share.

Chapter 1

1.1 See Paul Halpern's 'Pursuit of Destiny' and Peter Bernstein's 'Against the Gods', a book on the history of risk assessment

1.2 Raby, C. R., et al. (2007). "Planning for the future by western scrub-jays". Nature 445: 919 and at http://neurophilosophy.wordpress.com/2007/02/22/birds-plan-for-the-future/ on 111222

1.3 Pollard et al., Nature 443, 167-172 (14 September 2006), "An RNA gene expressed during cortical development evolved rapidly in humans"

1.4 Science News 12/31/2011; "Biology's Big Bang had a long fuse"

Chapter 3

3.1
    (http://www.sciencemuseum.org.uk/exhibitions/energy/site/E
    IZCaseStudy30.asp as of 090504)

Chapter 4

4.1 http://www.guardian.co.uk/uk/2012/mar/01/couple-guilty-boy-murder-witchcraft on 120530)

4.2 http://www.thefreelibrary.com/Future+bright+for+fortunetell
ing+in+Vt.+town-a01611632400

4.3 http://www.answers.com/topic/fortune-teller

Chapter 6

6.1 Amos Tversky and Daniel Kahneman, "Evidential Impact of
Base Rates," in *Judgment Under Uncertainty: Heuristics and
Biases,* Cambridge University Press, 1982, pp. 153–160; and
noted in http://www.schneier.com/essay-155.html

6.2 (David Ropeik and George Gray's book *Risk: A Practical
Guide for Deciding What's Really Safe and What's Really
Dangerous in the World Around You*)

6.3 Note: actually, this is also involved in another interesting
aspect of prediction that we've discussed briefly, and that is
the ability (or at least the *perceived* ability) to have control
over the future. Virtually no one gets car-sick when he is the
driver, but only the passenger. Similarly fear of flying is
almost entirely due to having no direct control of the
airplane. People feel they can easily use sun-screen, or wear
protective clothing, even though often they forget or are just
too lazy to do so. Skin cancer still kills but most people feel
that there's "nothing we can do" about that nuclear plant next
door, or, gasp!, give up our cell phones?!

6.4 Benjamin Radford at http://www.mediamythmakers.com/
091003

Chapter 9

9.1 George Box, G.E.P., Robustness in the strategy of scientific
model building, in **Robustness in Statistics**, R.L. Launer and
G.N. Wilkinson, Editors. 1979, Academic Press: New York
(http://www.anecdote.com.au/archives/2006/01/all_models_a
re.html)

9.2 Cornell's Hod Lipson, Science News 1/14/2012 p.21

9.3
www.msci.com/resources/research/monthly/Research200801
00.pdf

9.4 Dated 3/5/12 at dailymail.co/news/article-2023514/John-R-
Ginther-won-lottery-4-times-Stanford-University-statistics-
PhD.html)

9.5 Radvansky, G.A., & Copeland, D. E. (2006). "Walking through
doorways causes forgetting: Situation models and
experienced space", Memory & Cognition, 34, 1150

9.6 Science Writers conference Oct 17, 2011, as described on the
60-Second-Science podcast on Oct. 19, 2011

9.7 Statements like this really beg the question "How does the
brain *do* this, 'make a decision?' I have purposely avoided
discussions of neuro-anatomy. Although I feel that subject is
worthwhile, I personally feel that the ideas and concepts
discussed here are parallel to **where** the neuro-processing
takes place. Similarly, I have minimized discussions of
network theory, decisions trees, truth tables etc. Exactly *how*
does the brain 'make a decision?' That's a great subject for
further theory and experiment. A reasonable scenario for me
would be that, as the brain is building the various models for
potential outcomes for the current situation, it is also
'running them through' comparisons of the stored levels of
good and bad outcomes of the memory/patterns it is
evaluating. This repetitive "cycling through previous
patterns" is consistent with the proposal from Crick & Koch
(see their commentary "A Framework for Consciousness"
http://www.klab.caltech.edu/Papers/438.pdf) where they
suggest that this cycling is how the brain *selects* a relevant
and appropriate pattern. Some of these previous experiences
had good outcomes, and some had not so good outcomes.
All of these stored results (strengths of the connections in
these networks/patterns) could come together for comparison
of the previously stored results (even the summation of newly
associated results from the process we call imagination).

133

Perhaps they are compared one-to-one with the best result remaining until either the comparisons are done, or other external factors cause the brain to "just decide already!" Call that jumping to a conclusion. Or perhaps the brain can't easily determine the best choice. Call that analysis-paralysis or worse … too late, the lion just pounced.

9.8 Some fascinating additional research on the neurobiology of emotions is covered in the 1/13/2010 Brain Science podcast with Jaak Panksepp, Northwestern U. and his book "Affective Neuroscience"

9.9 There are other mechanisms that the brain uses to process information. These include not only the combination of multiple neurons and their firings, but also multiple firings in a short space of time. Some excellent discussion of these different mechanisms can be found in the book by Bialek, van Steveninck, Rieke, and Warland, "Spikes: Exploring the Neural Code"

9.10 Lookup IQ, and particularly Cattell-Horn-Carroll theory

Chapter 11

11.1 Karl Popper, "Objective Knowledge: An Evolutionary Approach" 1972

11.2 Quoted by PZ Myers on All in the Mind podcast 3/27/10

11.3 This phrase, 'As any schoolboy knows' pays homage to one of the first scientific books I ever read. "The Brain" by C.U.M. Smith (1970) where he says (p.46) "Every schoolboy mathematician will recognize that [the integral of $1/x = \ln(x)$]". A phrase that stopped me in my tracks, not yet having taken calculus. In fact, even after having taken calculus, it never was clear why this was so, but just that it was so.

11.4 As discussed by Michael Shermer on Point of Inquiry podcast, 6/6/11

11.5 Stephen J. Gould In an 1997 essay "Non-Overlapping Magisteria" for Natural History magazine, and later in his book *Rocks of Ages* (1999)

Chapter 12

12.1 Science News Feb 11 2012, p.28

12.2 At LiveScience.com (http://www.livescience.com/health/050808_human_conscio usness.html).

12.3 Dr's Doidge and Sachdev, Brain Science podcast 9/23/2010

12.4 http://www.newscientist.com/article/dn16809-body-illusions-rubber-hand-illusion.html

12.5 http://www.livescience.com/6917-brain-breakthrough-scientists-ll.html

12.6 Brain Science Podcast 2/28/12

12.7 Brain Science podcast 10/9/2009

12.8 60 Second Science podcast 5/21/10

12.9 Bill Nye on Point of Inquiry podcast 11/7/11

12.10 Nova Science Now podcast 3/14/11

12.11 Hugo Mercier, Point of Inquiry podcast 8/15/11

12.12 Robert Axelrod, the professor who recommends the tit-for-tat strategy from his game-theory competitions, also notes that your perception of the likelihood you will meet someone again [or someone very similar] affects how you will act in the new encounter. How likely it is that someone will 'defect' and not adopt the 'be nice' attitude. In his best professorial prose, Axelrod describes it like this (p. 15 in "The problem of Cooperation") "The principle is based on the weight of the next move, relative to the current move, to be sufficiently large, to make the future important. In other words, the discount parameter 'w' must be large enough to make the future loom large in the calculation of the total payoffs. After all, if you're unlikely to meet the other person again, or if you care little about future payoffs, then you might as well defect now and not worry about the consequences for the future."

12.13 Also described as 'contingent cooperation.' A nice article on these concepts, "Furry Friends Forever" by Susan Galdos in the April 7, 2012 issue of Science News

12.14 This is a widely used experiment in social-economics called the 'ultimatum game.'

12.15 http://www.predictioneersgame.com/

12.16 http://www.good.is/post/the-new-nostradamus/

Appendix 1 : The Statistics of Being a Successful Medium

Below is what I would call a 'non-rational' way to analyze the results of a medium's prognostications. What I mean by that is rather than set up a series of logical tests (the way most skeptics would) I suggest that 'regular' people react differently when they are told of the outcomes of a medium's predictions.

Looking at what happens in somewhat of a complete way, there are three things that occur. First a medium makes a prediction as to whether something good will happen, or something bad will happen. Next, the medium might offer (for a modest sum) a recommended course of action, especially if bad news is approaching. There may also be a sequence of actions that need to be taken even for good news to come to pass. Lastly, we actually see what really happened, whether or not the medium's prediction came true.

So, we have (1) good or bad prediction [two possibilities], (2) took the recommended actions or not [two possibilities], and (3) good or bad or 'nothing' actually happened [three possibilities]. So in all there are twelve possible outcomes. These outcomes are all listed in the figure below, and are also repeated in a table format after that, in case you are more of a spreadsheet kind of person.

The figure and table show (in my opinion) how most people would react to each of these outcomes. For example, in looking at the first upper-left square of the figure, the medium predicted something bad was going to occur, and indeed, something bad actually happened. This square is in the larger box of 'Took action recommended'. While bad happened, <u>even though</u> the recommended action was taken, the end result would likely be that the medium would be seen as more authentic; that there was actually a real connection between the prediction and the outcome. Most likely either we would believe we didn't follow the recommendations precisely enough, or the medium would make that claim, but the end result is that bad was predicted, and bad happened. Their credibility rises, and a stronger 'cause & effect' is seen. If you go through the other 11 scenarios, as can be seen in the chart and table below, you will see that in 10 out of the 12 scenarios the medium is either neutral or seen as possible. In only 2 of the scenarios do we conclude that they are probably quacks.

This is only a 17% lose rate! No wonder séances, astrology, palm reading etc have been around for centuries?

Took action recommended — Bad happens / Good happens

	Predicted bad	Predicted good
Bad happens	They're for real! Let's go back and pay them more!	They're a quack; now we have to deal with it
Good happens	Glad we did what they said, let's go back and pay them more!	They're for real! Let's go back and pay them more!

Took no action — Bad happens / Good happens

	Predicted bad	Predicted good
Bad happens	They're for real! Let's go back and pay them more!	They're for real! Let's go back and pay them more!
Good happens	OMG we were so lucky. No negatives for anyhone	They're for back and pay them more!

Cause & effect strengthened

Took action recommended — Nothing happens

	Predicted bad	Predicted good
Nothing happens	They may have been right, but we got lucky!	They're a quack; now we have to deal with it

Took no action — Nothing happens

	Predicted bad	Predicted good
Nothing happens	OMG we were so lucky. No negatives for anyhone	Hmm, I wonder what would have happened?

also in a table format ….

Prediction	Action	Result	Response	C&E strengthened?
Predict bad	Took action recommended	Bad happened	Ouch! But, it's for real!	Yes
Predict bad	Took action recommended	Good happened	omg! Go back for more!	Yes
Predict bad	Took action recommended	(nothing happened)	We were saved!	Yes
Predict bad	Took no action	Bad happened	Shoulda listened!	Yes
Predict bad	Took no action	Good happened	wow let's celebrate!	neutral
Predict bad	Took no action	(nothing happened)	oh well, let's eat	neutral
Predict good	Took action recommended	Bad happened	They're quacks!	No
Predict good	Took action recommended	Good happened	It's for real!	Yes
Predict good	Took action recommended	(nothing happened)	They're quacks!	No
Predict good	Took no action	Bad happened	I got punished	Yes
Predict good	Took no action	Good happened	Did they really know?	neutral/Yes
Predict good	Took no action	(nothing happened)	oh well, what's on tv?	neutral

139

Appendix 2 : The Justice System and Contra-causal Free Will

It is interesting to ponder the purpose of laws. Are they not there to restrict someone from doing what they would do otherwise, and impose penalties or punishments if they go ahead and 'do so anyway'? In the main text, as we discussed the strategy of 'tit for tat' (*tft*), Robert Axelrod also noted an important other strategy which was to defect first and then move on; always taking, never giving. This is the tactic of opportunists and, in the extreme, of bullies and dictators. If the general tactic of *tft* is the best way for groups to mutually coexist, then what is done with those who seek to take advantage? Laws and punishments are the solutions that societies have developed.

But the tricky part comes when someone is accused of wrong-doing. Did they truly break a law and so should face the consequences? Is there another opportunist who is falsely accusing someone for their own benefit, and so breaking other laws in doing so? To help with these conundrums, societies have developed a system of courts and judges/juries. And in developing these social tools, societies still struggle with the most fundamental issue during deliberations ... not only what happened, but more importantly what was the ***intent*** of those involved? There are the fine lines of intentional vs negligent behavior, and they both carry their consequences; but it is almost universal that if there was no intent of harm, or wrong-doing, then the punishments are, at least, very reduced.

There have been many discussions on the distinctions between *justice* and *punishment/retribution* in these efforts of societies. Perhaps no more so than when the litigants begin to introduce the ideas of mental impairment and claims of permanent or temporary insanity. These defenses seem to appear frequently in the headlines of high-profile cases. Society wrestles with these concepts because there doesn't seem to be a clear way to tease apart the ideas of blame, responsibility, intent, harm, damage, justice and fairness. How do we, as a society, appropriately deal with someone who is clearly mentally ill? (And I won't get into the discussion here, about how to measure/judge whether someone is clearly mentally ill; today it seems it still comes down to the opinion of experts, who often disagree depending on which side of the defense/prosecution they are on.) For those who need help, society seems to feel the best solution is to place them in a treatment environment. This accomplishes two things, first it (hopefully?) helps them become more well. A fascinating discussion on this was had on an All in the Mind Podcast, 11/25/2010. In that podcast they talked about a concept called 'Nido therapy' (nest therapy) where the emphasis is not on how to help patients function better in the broader society, but rather how to make them more comfortable and secure in their *own* environment. This is done by modifying their environment to better suit their psychological issues. A second benefit is that it helps society **predict** the behavior of the patient (now patient, rather than criminal?) in that they are now removed from society in general, and so society will not need to try to predict the patient's

141

behavior in future interactions. And it is this 'predict the behavior' of criminals that seems to be the underlying desire in courts' decisions.

If we focus on society's ability to confidently predict the future behavior of those involved, a new approach emerges. Rather than simply relate 'intent' with the outcomes and/or degree of punishment, the real focus should be on 'how much can we rely on this person to behave properly in the future?' I would suggest that this is actually the way societies proceed through their trials anyway, even though it is not explicitly thought of in that manner.

Do we believe them? Are they credible? Are they responsible? All of these questions are really addressing the question "can we trust them in the future?" For most criminals the solution seems to be to remove their freedom, and perhaps even provide some kind of social/psychological counseling with the expectations (hopes?) that bad people will behave better if/when released. But what about those whose mental capabilities are part of the reason their behavior was unacceptable in the first place? What about the distinction that courts wrestle with a lot lately, between mentally impaired and "temporary insanity"; how do we deal with these quagmires? How can we predict behavior if people don't have true control over their purposeful actions? Do they have free will or not?

I have listened to quite a few discussions on the subject of free will as to whether or not it exists and Sam Harris' book on the subject covers a great deal of ground. Some of the more compelling

142

arguments that free will **does** exist come from the hard sciences. Some proponents have invoked quantum mechanics and the uncertainties it demands, but we mustn't forget that QM is itself a mathematical model. Perhaps just as relevant is the idea of Brownian motion; this is the observation that at the microscopic level everything one can see is in constant movement. This was explained (i.e. described mathematically) by an obscure scientist in 1905 at the same time he offered up a couple of other tidbits of science.

What do these mean? For our discussions here, talking about human behavior, intent etc. how does the constant, random movement of molecules affect things? Philosophers have suggested that *if* identical conditions were to arise, identical to the conditions that were present when we made 'decision A', then the next time those identical conditions arose, we would have **no choice** but to make the same decision A; i.e. there would be no free will. As it turns out, the way our neurons work is to 'prime' themselves to get ready to fire if an adequate signal (or set of signals) arrive from other neurons (or in the case of sensory neurons, from our environment). They do this by spending energy to separate charges across the neural membrane (something like charging up a small battery). When the necessary signal arrives, the nerve cell opens ion-channel-gates in its membrane and allows this charge-separation to equalize across the membrane (kind of like opening the flood gates in a canal). It's this change that runs along the nerve fiber, down to the synaptic terminal to send a signal to the next nerve in the   network. Even this step is accomplished by molecules (neurotransmitters) randomly wandering in the space between the upstream and downstream nerves until they contact that downstream nerve which detects them as 'the signal' being sent.

Whether or not the nerve fires is dependent not only on the number of connections to other nerve cells, but also all of these molecules in the cell surroundings. This is why your electrolyte balance is so important. The chemicals that make up this flow across the membrane are primarily sodium, chlorine and calcium. If the concentrations of these chemicals change by even a small amount, nerve firing is impacted. A physical consequence could be ... you faint. While you can control your intake of both fluids and salts to generally maintain these concentrations, the details of these balances and these movements are all demonstrably **random**, both in theory and by observation. So, after all of this gibberish, the bottom line is that it is definitely impossible to exactly match any previous conditions that would be needed to test the theory of 'contra-causal free will'. This concept claims that *if* the exact same conditions were met, you would take the exact same path. But it's a false premise, since the *exact* same conditions can never be met. It's a useless 'if' ... or more in the context of this discussion, it's an idea that has no predictive value. If I could peer before the beginning of time, I could see .... Whatever ... it's meaningless. Archimedes practiced hyperbole when he claimed, "Give me a place to stand and I can move the earth." The river will never be the same the second time we step in it, and our mental biochemistry will never   be exactly the same either. Free will exists, and the result of that is that people, and their brains, have the capacity to perceive the environment and **make decisions** based on their memory patterns and their own models of how the future will unfold. The brain chooses actions that it expects will result in an improved environment for the individual and his or her groups (see the discussion on YESIR in the main text). If the actions they decide to take result in damage to others in society, then the laws are directed at them. If we, as a society, cannot determine (that's the hard part) how predictable someone is likely to be in future behavior, we remove them from society. We effectively remove them from our calculations about our *own* future. Today we seem to rely on a short list of punishments such as fines, loss of freedom, and even death, as tools to try to change the future behaviors of criminals. It appears that education and psychological counseling have been tried, and continue to be so, but generally have not met with the results that society seems to hope for. You might predict, from the content of this book, that my bias is that this education and counseling is the right way to go, and you'd be right. It seems the problem is that this education and counseling is more expensive than simply keeping someone in a small enclosure for decades, or just ending his life outright. These are the decisions and choices that we, as a society, have made ... and we are still learning the outcome of these predictions on our own safety, and futures.

Made in the USA
Charleston, SC
08 June 2012